AF595112

Electrolysis Processes

Electrolysis Processes

Special Issue Editor

Tanja Vidakovic-Koch

MDPI • Basel • Beijing • Wuhan • Barcelona • Belgrade • Manchester • Tokyo • Cluj • Tianjin

Special Issue Editor
Tanja Vidakovic-Koch
Max Planck Institute for
Dynamics of Complex Technical
Systems Sandtorstraße
Germany

Editorial Office
MDPI
St. Alban-Anlage 66
4052 Basel, Switzerland

This is a reprint of articles from the Special Issue published online in the open access journal *Processes* (ISSN 2227-9717) (available at: https://www.mdpi.com/journal/processes/special_issues/electrolysis_processes).

For citation purposes, cite each article independently as indicated on the article page online and as indicated below:

LastName, A.A.; LastName, B.B.; LastName, C.C. Article Title. *Journal Name* **Year**, *Article Number*, Page Range.

ISBN 978-3-03936-386-5 (Pbk)
ISBN 978-3-03936-387-2 (PDF)

Contents

About the Special Issue Editor

Tanja Vidakovic-Koch studied chemical engineering at University of Belgrade, Serbia, where she also obtained her Magister of Science degree in electrochemistry. She pursued further doctoral studies in process and systems engineering at Max Planck Institute for Dynamics of Complex Technical Systems (MPI-DCTS), Magdeburg, Germany, where she obtained her Dr.-Ing. degree as well as lecturer (Habilitation) qualification for Electrochemical Process Engineering ("venia legendi") from Otto von Guericke University (OvGU) in Magdeburg, Germany. Currently she is Head of the Electrochemical Energy Conversion group at the MPI-DCTS, Magdeburg, Germany, and she lectures at OvGU. Her research focuses on the model-based analysis of linear and nonlinear dynamics of electrochemical energy conversion in polymer electrolyte membrane (PEM) fuel cells, PEM electrolyzers, and enzymatic fuel cells.

Editorial

Editorial on Special Issue Electrolysis Processes

Tanja Vidaković-Koch

Electrochemical Energy Conversion, Max Planck Institute for Dynamics of Complex Technical Systems, D-39106 Magdeburg, Germany; vidakovic@mpi-magdeburg.mpg.de

Received: 7 May 2020; Accepted: 7 May 2020; Published: 14 May 2020

Renewable energies such as solar, hydro or wind power are in principal abundant but subjected to strong fluctuations. Therefore, development of new technologies for storage of these renewable energies is of special interest. Electrochemical technologies are ideal candidates for the use of excess current, and consequently an increased electrification of chemical processes is expected. In this respect, there are different pathways to utilize excess current electrochemically. Intermediate energy storages in (a) chemical energy carriers like hydrogen via water electrolysis or (b) electrochemical energy storage devices like batteries are perhaps most accepted and discussed solutions. Additionally, excess current can be used with the main goal not to be stored for later use, but to solve some environmental issue or for construction purposes. Possible applications are waste water treatment and electromachining. The article collection in this special issue consists of one review paper and nine original research papers and discusses these topics in more detail. As a Guest Editor of this special issue, I thank all authors for their contributions and wish all readers interesting insights and new inspirations for their works.

Electrolysis Processes for Intermediate Energy Storage in Chemical Energy Carriers Like Hydrogen–Water Electrolysis

There are three main technologies for water electrolysis: alkaline water electrolysis (AEL), proton exchange membrane (or polymer electrolyte membrane) electrolysis (PEMEL), and solid oxide electrolysis (SOEL), with two of them (AEL and PEMEL) at high technical readiness level. Despite these facts and intensive discussions on water electrolysis as a key technology for generation of pure hydrogen using renewable electricity, currently less than 4% of hydrogen originates from electrolysis, with the main part originating not from water but from from chlor-alkali electrolysis where hydrogen is a by-product of chlorine production [1]. Broad introduction of cost competitive and preferably zero carbon routes for hydrogen production is urgently required. In the review paper of Brauns and Turek [1], some of the main challenges hindering broader penetration of water electrolysis are discussed with a focus on AEL. As the authors write, AEL is a key technology for large-scale hydrogen production powered by renewable energy. However, conventional electrolyzers are designed for operation at fixed process conditions, therefore, the implementation of fluctuating and highly intermittent renewable energy is challenging. Their system analysis enabled important insights and a roadmap for more energy efficient systems. According to these authors, in order to be competitive with the conventional path based on fossil energy sources, each component of a hydrogen energy system needs to be optimized to increase the operation time and system efficiency. They stress that by combining AEL with hydrogen storage tanks and fuel cells, power grid stabilization can be achieved. As a consequence, the conventional spinning reserve can be reduced, which additionally lowers the carbon dioxide emissions.

Water electrolysis was also in the focus of Vorhauer et al. [2] and Altaf et al. [3]. One of the reasons for lowering energy efficiency of water electrolysis at high current densities is mass transport limitation in porous transport layers (PTL) of water electrolyzers. This issue is intrinsic for both low temperature water technologies (AEL and PEMEL) and is likely caused by the counter current transport of oxygen gas produced at the anode to the educt (water or OH- ions). In two publications by Vorhauer et al. [2] and Altaf et al. [3], this issue was studied with the help of porous network theory and with special emphasis on PEMEL. Pore network models (PNM) are powerful tools to simulate invasion and transport processes in different porous media with applications across different disciplines like geology, chemical engineering as well as electrochemical engineering (e.g., fuel cell applications). In their pioneering contribution, Vorhauer et al. [2] described a first application of a PNM of drainage for the prediction of the oxygen and water invasion process inside the anodic PTL at high current densities. According to the authors, in the simulation with narrow pore size distribution, the volumetric ratio of the liquid transporting clusters connected between the catalyst layer and the water supply channel is only around 3% of the total liquid volume contained inside the pore network at the moment when the water supply route through the pore network is interrupted; whereas around 40% of the volume is occupied by the continuous gas phase. The majority of liquid clusters are disconnected from the water supply routes through the pore network if liquid films along the walls of the porous transport layer are disregarded. Moreover, these clusters hinder the countercurrent oxygen transport. They also based on these sketches a new route for the extraction of transport parameters from Monte Carlo simulations, incorporating pore scale flow computations and Darcy flow.

In the publication by Altaf et al. [3], results from Monte Carlo pore network simulations are shown and compared qualitatively to microfluidic experiments from literature. The literature-based experimental invasion patterns of different types of PTLs, such as felt, foam or sintered, have shown good agreement with pore network simulations. Additionally, the impact of pore size distribution on the phase patterns of oxygen and water inside the pore network was studied. These very promising results further supported pore network modeling as a valuable tool for gaining new insights in the transport processes in porous layers during water electrolysis.

Reliable models of gas-liquid equilibria in aqueous electrolytes are of significant importance for proper description of many electrochemical processes including gases as products or reactants (major examples are water, brine, CO_2 electrolysis, but also reactions in polymer electrolyte fuel cells). Nabipour et al. [4] proposed a novel solubility estimation tool for gases in aqueous electrolyte solutions based on an extreme learning machine (ELM) algorithm. Although the presented study cases are not directly relevant for water electrolysis, the here developed novel methodology has a potential to be transferred to more relevant examples for electrochemical applications.

Electrolysis Processes for Intermediate Energy Storage in Electrochemical Energy Storage Devices

Different types of electrochemical energy storage devices are discussed in the framework of intermediate energy storage from renewables. Common examples are rechargeable batteries like Li-ion or redox flow batteries. In this special issue, two less discussed options are described, a so-called acid-base flow battery [5] and supercapacitors [6]. An acid-base flow battery is proposed by Xia et al. [5]. This very interesting solution is based on reverse electrodialalysis with bi-polar membranes. During charging, the system operates in electrolysis mode; hydrogen and oxygen evolution reactions take place at the cathode and anode, respectively; and acid and base solutions are regenerated in corresponding compartments. Alike normal water electrolysis, the two half-reactions take place at different pH values (acidic and alkaline conditions). During discharge, neutralization of acid and base produces electricity in the process of reverse electrodialysis with bipolar membranes in an apparently fairly overlooked flow battery concept. The authors demonstrated this concept with stack experiments, consisting of 5 to 20 repeating cell units at lab scale. The first results are very promising and the studied system shows already energy density in the similar range of redox flow batteries. The challenges and measures to further increase energy efficiency of this new type of acid-base flow battery are discussed.

Due to the high power fluctuations that are inherent to renewable energy sources, the dynamics of the storage media is of great importance when designing storage concepts for renewable energy. Electrochemical storage systems showing even better dynamics than batteries are so-called supercapacitors. These devices can be quickly charged and discharged, but have lower energy density than batteries. Still, due to their favorable dynamics, they can be a valuable addition to batteries in the framework of intermediate energy storage from renewables. The storage capacity of these devices depends largely on employed materials. In this special issue, development of novel electrode material for supercapacitor application based on pseudocapacitance is discussed. Guragain et al. [6] developed a large-surface-area MCO_2O_4 material in which a tubular microstructure leads to a noticeable pseudocapacitive property with the excellent specific capacitance value exceeding 407.2 F/g at 2 mV/s scan rate. In addition, a Coulombic efficiency ~100% and excellent cycling stability with 100% capacitance retention was noted even after 5000 cycles. These tubular MCO_2O_4 microstructures display peak power density exceeding 7000 W/kg. Based on these authors, the superior performance of the tubular MCO_2O_4 microstructure electrode is attributed to their high surface area, adequate pore volume distribution, and active carbon matrix, which allows effective redox reaction and diffusion of hydrated ions.

Electrolysis Processes Aiming to Solve Environmental Issues or for Construction Purposes

In addition to clean energy, the requirement of clean water is of the upmost importance for prosperity and healthiness of the world population. With respect to water, one issue is water scarcity, causing 1.2 billion people to suffer from a lack of water [7]. Another issue is industry or domestic-based water pollution [7–9]. Due to the widespread use of chlorinated compounds such as HCl, NaCl, and $MgCl_2$ in the industry, the content of chloride ion in wastewater is increasing [8,9]. If it is discharged directly to the environment, diverse environmental issues will be created with consequences for water resources, soil and human health. At present the most widely used method for Cl^- waste water treatment is chemical precipitation. Electrochemical oxidation processes have a high potential for efficient removal of even trace amounts of Cl^- due to high degradation efficiency, no secondary pollution, modularity, flexibility, and use of renewable electricity, which can contribute to stabilization of the energy grid. The degradation efficiency and degradation products of the electrochemical oxidation process change with the anode material, therefore, two articles by Xu et al. [8] and Liu et al. [9] discuss the development of new anode materials for electrochemical Cl^- removal. Xu et al. [8] studied porous Ti/SnO_2-Sb_2O_3/PbO_2 electrodes for electrocatalytic oxidation of chloride ions by varying different parameters. The results have shown that Ti/SnO_2-Sb_2O_3/PbO_2 electrodes with 150 μm pore size had the best removal effect on chloride ions with removal ratios amounting up to 98.5% when the initial concentration of chloride ion was 10 g L^{-1}. The advanced electrode structure minimized oxygen evolution as a side reaction, which further increased the removal effect of chloride ions. In the further publication, Liu et al. [9] studied different material combinations. The porous Ti/Sb–SnO_2/Ni–Ce–PbO_2 electrode was prepared by using a porous Ti plate as a substrate, an Sb-doped SnO_2 as an intermediate, and a PbO_2 doped with Ni and Ce as an active layer. The authors studied also the mechanism of electrochemical dechlorinating. They found out that the increase of OH^- inhibits the degradation of Cl^-.

In addition to inorganic catalysts, oxidation of mainly organic pollutants in the waste water treatment can be performed with the help of biological catalysts, for example microorganisms. These systems can operate either as microbial fuel cells (MFC) or in electrolysis mode. A contribution by Muddemann et al. [7] discuss current challenges in the scale up of these systems, with special emphasis on oxygen reducing cathode. The authors demonstrated a strong increase in the MFC performance and long term stability upon improving catalyst coating quality.

Finally Liu et al. [10] describe electrochemical discharge machining (ECDM) as an effective way to fabricate micro structures in non-conductive materials, such as quartz glass and ceramics. This has significant importance for the development of micro electromechanical systems (MEMS), such as micro reactors and micro medical devices, which often consist of the micro structures of nonconductive materials, such as glass, ceramics and silicon nitride and which are difficult to fabricate using the traditional machining methods.

Funding: This research received no external funding.

Conflicts of Interest: The author declares no conflict of interest.

References

1. Brauns, J.; Turek, T. Alkaline water electrolysis powered by renewable energy: A review. *Processes* **2020**, *8*, 248. [CrossRef]
2. Vorhauer, N.; Altaf, H.; Tsotsas, E.; Vidakovic-Koch, T. Pore network simulation of gas-liquid distribution in porous transport layers. *Processes* **2019**, *7*, 558. [CrossRef]
3. Altaf, H.; Vorhauer, N.; Tsotsas, E.; Vidaković-Koch, T. Steady-state water drainage by oxygen in anodic porous transport layer of electrolyzers: A 2D pore network study. *Processes* **2020**, *8*, 362. [CrossRef]
4. Nabipour, N.; Mosavi, A.; Baghban, A.; Shamshirband, S.; Felde, I. Extreme learning machine-based model for solubility estimation of hydrocarbon gases in electrolyte solutions. *Processes* **2020**, *8*, 92. [CrossRef]
5. Xia, J.; Eigenberger, G.; Strathmann, H.; Nieken, U. Acid-base flow battery, based on reverse electrodialysis with Bi-polar membranes: Stack experiments. *Processes* **2020**, *8*, 99. [CrossRef]
6. Guragain, D.; Zequine, C.; Gupta, R.; Mishra, S. Facile synthesis of bio-template tubular MCo_2O_4 (M = Cr, Mn, Ni) microstructure and its electrochemical performance in aqueous electrolyte. *Processes* **2020**, *8*, 343. [CrossRef]
7. Muddemann, T.; Haupt, D.; Jiang, B.; Sievers, M.; Kunz, U. Investigation and improvement of scalable oxygen reducing cathodes for microbial fuel cells by spray coating. *Processes* **2020**, *8*, 11. [CrossRef]
8. Xu, K.; Peng, J.; Chen, P.; Gu, W.; Luo, Y.; Yu, P. Preparation and characterization of porous $Ti/SnO_2–Sb_2O_3/PbO_2$ electrodes for the removal of chloride ions in water. *Processes* **2019**, *7*, 762. [CrossRef]
9. Liu, S.; Gui, L.; Peng, R.; Yu, P. A novel porous Ni, Ce-doped PbO_2 electrode for efficient treatment of chloride ion in wastewater. *Processes* **2020**, *8*, 466. [CrossRef]
10. Liu, Y.; Zhang, C.; Li, S.; Guo, C.; Wei, Z. Experimental study of micro electrochemical discharge machining of ultra-clear glass with a rotating helical tool. *Processes* **2019**, *7*, 195. [CrossRef]

Review

Alkaline Water Electrolysis Powered by Renewable Energy: A Review

Jörn Brauns * and Thomas Turek

Institute of Chemical and Electrochemical Process Engineering, Clausthal University of Technology, Leibnizstr. 17, 38678 Clausthal-Zellerfeld, Germany; turek@icvt.tu-clausthal.de
* Correspondence: brauns@icvt.tu-clausthal.de; Tel.: +49-5323-72-2473

Received: 23 January 2020; Accepted: 18 February 2020; Published: 21 February 2020

Abstract: Alkaline water electrolysis is a key technology for large-scale hydrogen production powered by renewable energy. As conventional electrolyzers are designed for operation at fixed process conditions, the implementation of fluctuating and highly intermittent renewable energy is challenging. This contribution shows the recent state of system descriptions for alkaline water electrolysis and renewable energies, such as solar and wind power. Each component of a hydrogen energy system needs to be optimized to increase the operation time and system efficiency. Only in this way can hydrogen produced by electrolysis processes be competitive with the conventional path based on fossil energy sources. Conventional alkaline water electrolyzers show a limited part-load range due to an increased gas impurity at low power availability. As explosive mixtures of hydrogen and oxygen must be prevented, a safety shutdown is performed when reaching specific gas contamination. Furthermore, the cell voltage should be optimized to maintain a high efficiency. While photovoltaic panels can be directly coupled to alkaline water electrolyzers, wind turbines require suitable converters with additional losses. By combining alkaline water electrolysis with hydrogen storage tanks and fuel cells, power grid stabilization can be performed. As a consequence, the conventional spinning reserve can be reduced, which additionally lowers the carbon dioxide emissions.

Keywords: alkaline water electrolysis; hydrogen; renewable energy; sustainable; dynamic; fluctuations; wind; solar; photovoltaic; limitations

---

1. Introduction

Hydrogen is considered a promising energy carrier for a sustainable future when it is produced by utilizing renewable energy [1]. Today, less than 4% of hydrogen production is based on electrolysis processes, of which the main part is hydrogen as a by-product of chlorine production. Hence, the major share of the needed hydrogen depends on the fossil path through the steam reforming of natural gas [2]. This situation is caused by the higher production costs of electrolysis processes compared to the conventional fossil sources, due to high electricity costs and interfering laws [3]. To reduce CO_2 emissions and to become independent of fossil energy carriers, the share of hydrogen produced using renewable power sources needs to be increased significantly in the next few decades. Therefore, water electrolysis is a key technology for splitting water into hydrogen and oxygen by using renewable energy. After drying and removing oxygen impurities, the hydrogen purity is higher than 99.9%, and the hydrogen can be directly used in the following processes or in the transport sector [4]. Solar and wind energy are the preferred renewable power sources for hydrogen production, as their distribution is the most widespread [5,6]. Hydropower, biomass, and geothermal energy are alternatives, and are often utilized for the base load [7]. The main problem with using renewable energy is the unevenly distributed and intermittent local availability [6]. With a higher share of renewable energy from wind turbines or solar photovoltaic panels and fair CO_2 emission costs, hydrogen production by water electrolysis will become more attractive. The combination of water electrolysis with renewable energy is particularly advantageous, as excess electrical energy can be chemically stored in hydrogen to balance the discrepancy between energy demand and production [6]. For large-scale applications, the hydrogen can be stored in salt caverns, storage tanks, or the gas grid [8–12]. Smaller hydrogen quantities can also be stored in metal hydrides [13,14].

For water electrolysis, there are three technologies available: Alkaline water electrolysis (AEL), proton exchange membrane (or polymer electrolyte membrane) electrolysis (PEMEL), and solid oxide electrolysis (SOEL) [15–18]. While the low-temperature technologies, AEL and PEMEL, both provide high technology readiness levels, the high-temperature SOEL technology is still in the development stage [19]. Alkaline water electrolysis uses concentrated lye as an electrolyte and requires a gas-impermeable separator to prevent the product gases from mixing. The electrodes consist of non-noble metals like nickel with an electrocatalytic coating. PEMEL uses a humidified polymer membrane as the electrolyte and noble metals like platinum and iridium oxide as the electrocatalysts. Both technologies are operated at temperatures from 50 to 80 °C and allow operation pressures of up to 30 bar. The nominal stack efficiency of both technologies is around 70% [18,20]. SOEL is also known as high-temperature (HTEL) or steam electrolysis, as gaseous water is converted into hydrogen and oxygen at temperatures between 700 and 900 °C. Theoretically, stack efficiencies near 100% are possible due to positive thermodynamic effects on power consumption at higher temperatures. However, the increased thermal demand requires a suitable waste heat source from the chemical, metallurgical, or thermal power generation industry for economical operation. Moreover, the corrosive environment demands further material development [6,20,21]. As a consequence, SOEL provides only small stack capacities below 10 kW, compared to 6 MW for AEL and 2 MW for PEMEL [20]. Hence, the investment costs and the lifetime determine whether AEL or PEMEL is the most favorable system design for a large-scale application. Today, the investment costs for AEL are from 800 to 1500 € kW^{-1} and for PEMEL from 1400 to 2100 € kW^{-1}. Furthermore, the lifetime of alkaline water electrolyzers is higher and the annual maintenance costs are lower compared to a PEMEL system [15,20,22,23]. Often, PEMEL systems are preferred for dynamic operation due to the short start-up time and a broad load flexibility range. The shortcomings of AEL are gradually being overcome by further development [24]. Therefore, this review focuses on alkaline water electrolysis powered by renewable energy. To ensure safety and high efficiency, alkaline water electrolyzers must be optimized for dynamic operation. Hence, the process needs to be analyzed for how the dynamics will affect the system performance and what aspects should be considered when fluctuating renewable energy is used instead of a constant load [25]. Thus, this contribution shows model descriptions for alkaline water electrolysis, photovoltaic panels, and wind turbines to identify the limitations when combining all components into a hydrogen energy system. Furthermore, theoretical models can help to solve the existing problems using intelligent system design and suitable operation strategies.

This study mainly contains literature that was obtained with the keywords alkaline electrolyzer (or electrolyser or electrolysis) in combination with one of the following words: Renewable, sustainable, green, dynamic, fluctuation, intermittent, solar, photovoltaic, wind, and power to gas. For a broader overview, additional literature is also included. Figure 1 shows the number of annual publications that are listed in the Web of Science Database for the given keywords from 1990 to 2019. Additionally, the keyword alkaline is replaced by other water electrolysis technologies to show the share of technology-specific publications [26]. Around 2010, the number of annual publications started to increase persistently due to the discussion about the energy turnaround, especially in Germany and other European countries [9,27]. Furthermore, the topic is often discussed technology-independently, as the number of technology-specific publications is small compared to publications with unspecified water electrolysis technologies. While the low-temperature technologies, AEL and PEMEL, show an equal share of technology-specific publications, the high-temperature technology SOEL is mentioned less. This distribution reflects the recent considerations of which technology may be favored for sustainable hydrogen production. Particularly, alkaline water electrolysis is considered as the most reliable method for large-scale hydrogen production [5,21].

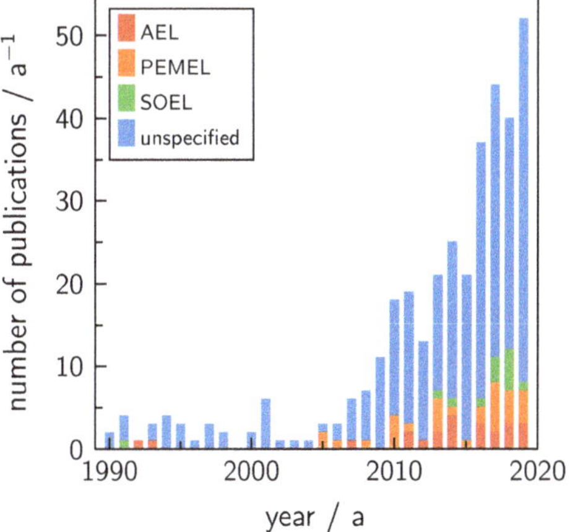

Figure 1. The number of publications per year from 1990 to 2019 containing the specified keywords. Around 2010, the publication rate increases due to greater interest in the energy turnaround. While the topic is often discussed technology-independently (unspecified), more publications for low-temperature technologies, like alkaline water electrolysis (AEL) and proton exchange membrane electrolysis (PEMEL), are available than for the high-temperature technology solid oxide electrolysis (SOEL) [26].

2. Alkaline Water Electrolysis

Alkaline water electrolysis is used to split water into the gases hydrogen and oxygen using electric energy. The chemical reactions are given in the Equations (1)–(3). At the cathode, water molecules are reduced by electrons to hydrogen and negatively charged hydroxide ions. At the anode, hydroxide ions are oxidized to oxygen and water while releasing electrons. Overall, a water molecule reacts to hydrogen and oxygen in the ratio of 2:1.

$$\text{Cathode:} \quad 2\,H_2O_{(l)} + 2\,e^- \longrightarrow H_{2(g)} + 2\,OH^-_{(aq)} \tag{1}$$

$$\text{Anode:} \quad 2\,OH^-_{(aq)} \longrightarrow 0.5\,O_{2(g)} + H_2O_{(l)} + 2\,e^- \tag{2}$$

$$\text{Overall reaction:} \quad H_2O_{(l)} \longrightarrow H_{2(g)} + 0.5\,O_{2(g)} \tag{3}$$

The required cell voltage for this electrochemical reaction can be determined by thermodynamics. The free reaction enthalpy $\Delta_R G$ in (4) can be calculated with the reaction enthalpy $\Delta_R H$, the temperature T, and the reaction entropy $\Delta_R S$.

$$\Delta_R G = \Delta_R H - T \cdot \Delta_R S \tag{4}$$

The reversible cell voltage U_{rev} in (5) is determined by the ratio of the free reaction enthalpy $\Delta_R G$ to the product of the number of exchanged electrons $z = 2$ and the Faraday constant F (96,485 C mol^{-1}) [28].

$$U_{rev} = -\frac{\Delta_R G}{z \cdot F} \tag{5}$$

At a temperature of 25 °C and an ambient pressure of 1 bar (standard conditions), the free reaction enthalpy for the water splitting reaction is $\Delta_R G = 237\,\text{kJ mol}^{-1}$, which leads to a reversible cell voltage of $U_{rev} = -1.23\,\text{V}$. As the free reaction enthalpy is positive at standard conditions, the water splitting is a non-spontaneous reaction [28]. Due to irreversibilities, the actual cell voltage needs to be higher than the reversible cell voltage for the water splitting reaction. The thermoneutral voltage U_{th} in Equation (6) depends on the reaction enthalpy $\Delta_{\text{R}}H$, which is composed of the free reaction enthalpy $\Delta_{\text{R}}G$ and irreversible thermal losses $T \cdot \Delta_{\text{R}}S$.

$$U_{\text{th}} = -\frac{\Delta_{\text{R}}H}{z \cdot F} \tag{6}$$

At standard conditions, the reaction enthalpy for water electrolysis is $\Delta_{\text{R}}H = 286\,\text{kJ}\,\text{mol}^{-1}$. Hence, the thermoneutral voltage is $U_{\text{th}} = -1.48\,\text{V}$ [28].

3. System

A schematic flow diagram of alkaline water electrolysis is shown in Figure 2. The electrolyte is pumped through the electrolysis stack, where the product gases are formed. While natural convection can be a cost-efficient alternative, gas coverage of the electrode surface can raise the required cell voltage and therefore increase the operational costs [29]. Additionally, most alkaline water electrolyzer systems provide a temperature control for the electrolyte to maintain an optimal temperature range.

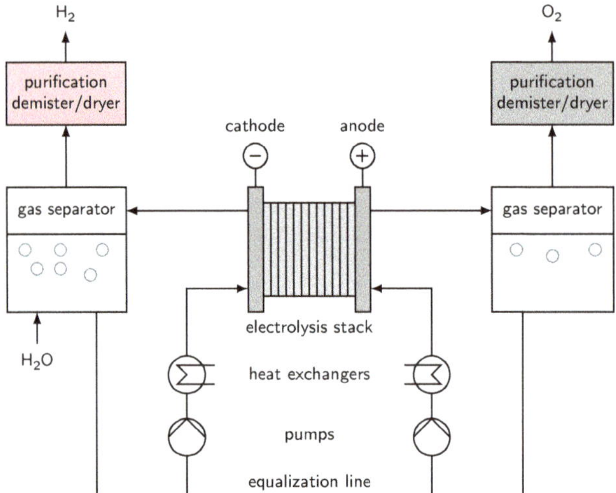

Figure 2. A schematic flow diagram of an alkaline water electrolyzer. The electrolyte is pumped through the electrolysis cell where the gas evolution takes place. Adjacent gas separators split both phases, and the liquid phase flows back to the electrolysis stack. Heat exchangers ensure that the optimal temperature is maintained, and the product gases can be purified afterward.

The two-phase mixtures of liquid electrolyte and product gas leave the electrolysis cell and enter subsequent gas separators. Mostly, the phase separation is realized with a high residence time in large tanks. The product gas is demisted and dried before it is purified to the desired level [30]. The liquid electrolyte leaves the gas separator and is pumped back to the electrolysis stack. As the product gases are soluble in the electrolyte solution, the mixing of both electrolyte cycles causes losses and higher gas impurities. An alternative can be to use partly separated electrolyte cycles with an equalization line for liquid level balancing of both vessels [31,32]. With separated electrolyte cycles, the electrolyte concentration will increase on the cathodic side due to water consumption and decrease on the anodic side due to water production. Therefore, the electrolyte requires mixing, on occasion, to maintain an optimal electrolyte conductivity.

4. Cell Design and Cell Voltage

The design of the electrolysis stack depends on the manufacturer; however, some general similarities can be observed. Two variants of cell designs are shown in Figure 3. Earlier alkaline water electrolyzers used a conventional assembly with a defined distance between both electrodes. Later, this concept was replaced by the zero-gap assembly, where the electrodes are directly pressed onto the separator to minimize ohmic losses due to the electrolyte. Porous materials like Zirfon™ Perl UTP 500 (AGFA) or dense anion exchange membranes can be used as the separator [33–37].

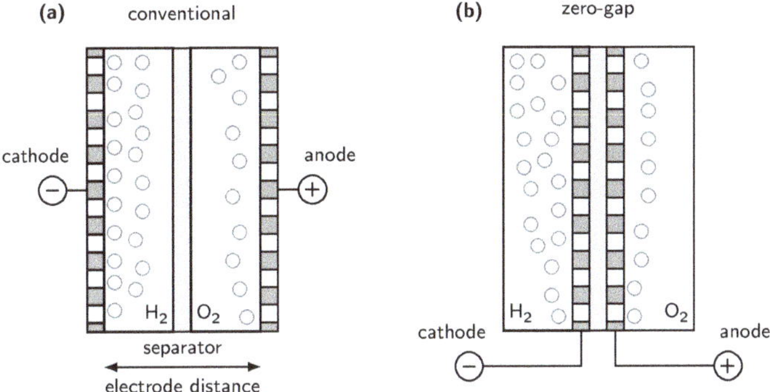

Figure 3. Different cell designs for alkaline water electrolysis. Whereas (**a**) shows a conventional assembly with a defined distance between both electrodes, (**b**) depicts a zero-gap assembly where the electrodes are directly pressed onto the separator [38].

During operation, the required cell voltage is always higher than the reversible cell voltage due to different effects. A calculated cell voltage profile is displayed in Figure 4. In addition to the ohmic losses, $I \cdot R_{ohm}$, there are activation overvoltages of the electrodes, η_{act}. The ohmic resistance of the cell design is affected by the electronic conductivity of the electrode material, the specific conductivity of the electrolyte, the ionic conductivity of the separator material, and gas bubble effects.

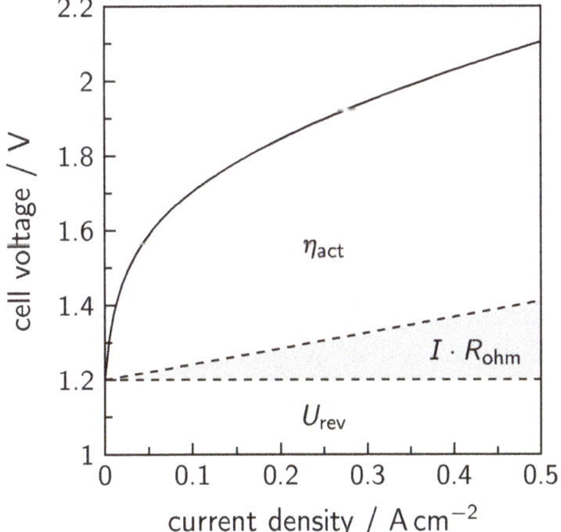

Figure 4. The calculated cell voltage of an atmospheric alkaline water electrolyzer at a temperature of 60 °C according to Equation (8). The overall cell voltage consists of the reversible cell voltage U_{rev}, ohmic losses $I \cdot R_{ohm}$, and activation overvoltages η_{act} [39,40].

The zero-gap design tries to eliminate the electrolyte losses by minimizing the electrode distance. There is still a minimal gap between both electrodes, which can increase the cell voltage. The activation overvoltages are defined by the electrode materials. Whereas nickel is the most-used electrode material, it provides very high overvoltages for the oxygen and hydrogen evolution reactions [41–44]. Hence, electrocatalytic materials are added to the electrodes. Iron is a cost-efficient catalyst for the oxygen evolution reaction [41,42,45]. Molybdenum decreases the overvoltage for the evolution of hydrogen at the cathode [44,46,47].

Several authors have proposed correlations for the modeling of cell voltage. Equation (7) considers the operation temperature ϑ and the current density j by describing the dependencies with empirical parameters. While the parameters r_i reflect ohmic losses, s and t_i stand for the activation overvoltages of the hydrogen and oxygen evolution reactions [28].

$$U_{\text{cell}} = U_{\text{rev}} + (r_1 + r_2 \cdot \vartheta) \cdot j + s \cdot \log\left[\left(t_1 + \frac{t_2}{\vartheta} + \frac{t_3}{\vartheta^2}\right) \cdot j + 1\right] \tag{7}$$

This correlation can be extended with the effects of the operation pressure p in (8) by adding the empirical parameters d_i, which specify the additional losses owing to pressurized operation [39]. In general, the reversible cell voltage increases with the pressure; however, the ohmic resistance caused by the gas bubbles decreases as the bubble diameter becomes smaller. Hence, both effects equalize each other and only small differences can be observed [48].

$$U_{\text{cell}} = U_{\text{rev}} + [(r_1 + d_1) + r_2 \cdot \vartheta + d_2 \cdot p] \cdot j + s \cdot \log\left[\left(t_1 + \frac{t_2}{\vartheta} + \frac{t_3}{\vartheta^2}\right) \cdot j + 1\right] \tag{8}$$

The correlations (7) and (8) are empirical and therefore only valid for the actual system to which they are adjusted. The correlation parameters and a suitable equation for the reversible cell voltage under atmospheric conditions can be found in the Appendix A in Table A1 and Equation (A1). Other authors have proposed physically reasonable models based on actual dimensions and properties of the system rather than on empirical correlations.

An example for such an approach is Equation (9), in which the terms are split into experimentally determinable parts [49].

$$U_{\text{cell}} = U_{\text{rev}} + \eta_{\text{act}}^{\text{c}} + \eta_{\text{act}}^{\text{a}} + I \cdot (R_{\text{c}} + R_{\text{a}} + R_{\text{ele}} + R_{\text{mem}}) \tag{9}$$

The cell voltage U_{cell} is calculated with the reversible voltage U_{rev}, the activation overvoltages η_{act}, and the ohmic resistances. Whereas R_{c} and R_{a} represent the reciprocal electronic conductivity of the electrode materials, R_{ele} stands for the ohmic loss caused by the electrolyte conductivity. Additionally, the ohmic resistance R_{mem} of the separator material is taken into account. The activation overvoltages η_{act} can be calculated with the Butler–Volmer equation. In most cases, the simplified Tafel equation is sufficient to describe the resulting overpotentials [40]. The required Tafel slope and exchange current density can be extracted from experimental data. Hence, those parameters are only valid for the actual system design; however, they can be easily replaced by other data when needed. As the ohmic resistances of the electrodes (R_{c} and R_{a}) only depend on the electronic conductivity and electrode dimensions, both values are known. In most cases, the ohmic resistance of the electrode is comparably small and can be neglected. The electrolyte resistance R_{ele} is determined by the specific electrolyte conductivity and the cell design. Whereas the electrolyte gap is minimal in zero-gap designs, conventional setups maintain a defined distance between both electrodes. As the specific conductivity of the electrolyte gap is affected by gas bubbles, there is an optimal electrode distance for conventional designs [50]. If the electrode distance is too small, the gas bubbles accumulate in the gap and lower the conductivity. With increasing distance, the bubble detachment is enhanced and the specific conductivity increases. It is a trade-off between a small electrolyte gap—as the ohmic resistance increases linearly with this parameter—and a better conductivity of the space between both electrodes.

In addition to the decreasing electrolyte conductivity with higher amounts of gas bubbles, the active electrode surface can be blocked by gaseous compounds, which leads to additional losses [49]. As this phenomenon depends on the cell design and operation concept, there are difficulties in describing it properly. Therefore, it is often neglected, or empirical correlations referring to the gas hold-up are utilized [49].

Furthermore, the installed separator material also has significant ohmic losses. While the porous separator Zirfon™ Perl UTP 500 is often used, anion exchange membranes are promising alternatives. For Zirfon™-based materials, experimental data of the resistance at a fixed electrolyte concentration for different temperatures are available [51].

The most-used electrolyte for alkaline water electrolysis is an aqueous solution of potassium hydroxide (KOH) with 20 to 30 wt.% KOH, as the specific conductivity is optimal at the typical temperature range from 50 to 80 °C [25]. A cheaper alternative would be a diluted sodium hydroxide solution (NaOH), which has a lower conductivity [52]. Calculated specific electrolyte conductivities for both electrolyte solutions at different temperatures are shown in Figure 5. While KOH provides a specific conductivity around $95\,\mathrm{S\,m^{-1}}$ at 50 °C, NaOH reaches a value around $65\,\mathrm{S\,m^{-1}}$. At a temperature of 25 °C, a similar effect can be seen. The conductivity of KOH is around 40 to 50% higher than the conductivity of a NaOH solution at the optimal weight percentage. Another aspect is the solubility of the product gases inside the electrolyte, as this influences the resulting product gas purity. In general, the gas solubility decreases with an increasing electrolyte concentration due to the salting-out behavior [53]. NaOH also shows a slightly higher salting-out effect than that of KOH. Hence, the product gas solubility is higher in a KOH solution [54–56].

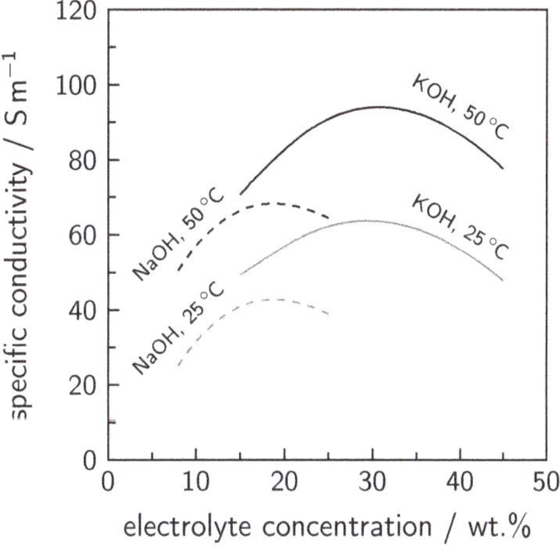

Figure 5. The calculated specific electrolyte conductivity as a function of the electrolyte concentrations of sodium hydroxide (NaOH) and potassium hydroxide (KOH) solutions at different temperatures obtained by Equations (A2) and (A3). The correlation parameters can be found in Table A2 [52,57].

Another approach is to use ionic liquids (ILs) as the electrolyte or as an additive, owing to their remarkable properties [5,6,21]. Ionic liquids are organic substances which are liquid at room temperature and are electrically conductive [58]. A negligible vapor pressure, non-inflammability, and thermal stability are promising arguments for their utilization in water electrolysis. Furthermore, ILs can be used over a wide electrochemical window [59]. The absorption and separation of gases is an additional area of application [60,61]. However, the toxicity of ILs is a current field of research, and the viscosity is comparably high, which should be taken into account before any large-scale implementation [6,58,59]. In addition to providing high-efficiency water electrolysis at low temperatures, ILs are chemically inert and therefore do not require expensive electrode materials [62].

5. Gas Purity

Gas purity is an important criterion of alkaline water electrolysis. While the produced hydrogen typically has a purity higher than 99.9 vol.% (without additional purification), the gas purity of oxygen is in the range of 99.0 to 99.5 vol.% [48]. As both product gases can form explosive mixtures in the range of approximately 4 to 96 vol.% of foreign gas contamination, technical safety limits for an emergency shutdown of the whole electrolyzer system are at a level of 2 vol.% [31,63]. Therefore, the product gas impurity needs to be below this limit during operation to ensure continuous production. Experimentally determined anodic gas impurities for alkaline water electrolysis are presented in Figure 6 for different operation modes. The current densities are in the range from 0.05 to 0.7 A cm^{-2} and the system pressures range from 1 to 20 bar [64].

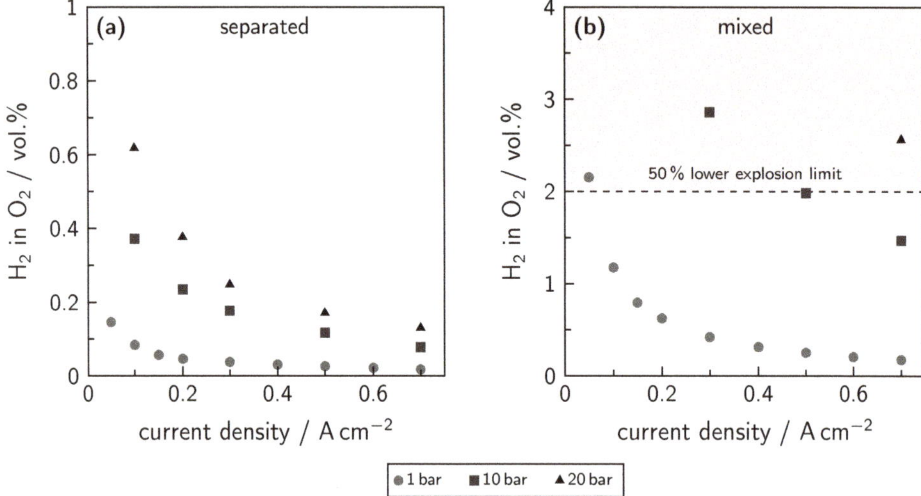

Figure 6. Anodic gas impurity (H_2 in O_2) in relation to the current density at different pressure levels for (**a**) separated and (**b**) mixed electrolyte cycles, at a temperature of 60 °C, with an electrolyte concentration of approximately 32 wt.% and an electrolyte volume flow of 0.35 L min^{-1} [64].

While the gas impurities with separated electrolyte cycles are below 0.7 vol.% for all tested current densities and pressure levels, mixing of the electrolyte cycles increases the gas impurity significantly. Furthermore, two similarities can be seen. The gas impurity lowers with increasing current density and increases at higher pressure levels. Both effects are physically explainable. While the contamination flux stays constant with varying current densities, the amount of produced gas becomes lower in a linear relationship. Hence, at a higher current density, the contamination is more diluted than at a lower current density [32,64]. As a consequence, the operation in the part-load range is more critical due to the higher gas impurity. The amount of dissolved product gas increases with pressure; thus, high concentration gradients for the diffusion through the separator material are available, and more dissolved foreign gas reaches the other half-cell when mixing [64]. However, operation at slightly elevated pressures is favorable, as the costly first mechanical compression level can be avoided by the direct compression inside the electrolyzer system [65]. With mixed electrolyte cycles, the gas impurity reaches critical values even at higher current densities during pressurized operation. While at atmospheric pressure, the gas impurity is only at a current density of 0.05 A cm^{-2}, slightly above the safety limit of 2 vol.% H_2 in O_2, this limit is already reached at 0.5 A cm^{-2} for a system pressure of 10 bar. At 20 bar, no sufficient gas purity could be measured, as even a current density of 0.7 A cm^{-2} results in a gas impurity of 2.5 vol.%.

6. Periphery

The operation of alkaline water electrolyzers is also affected by the installed periphery, including power supplies. The output signal may contain a specific number of ripples, which directly influences the process performance [66]. Power supplies with a high ripple propensity lower the overall efficiency and, therefore, signal smoothing is necessary. The ripple formation is avoidable, but the component costs will be higher [67]. In general, thyristor-based power supplies tend to deliver a higher degree of fluctuation, and transistor-based systems output a smoother signal. Additionally, a higher ripple frequency does not affect the system performance as much as the occurrence of low-frequency ripples [68]. Furthermore, a coherence between the ripple behavior of a power supply and the product gas quality of alkaline water electrolysis can be observed [69].

7. Renewable Energy

The combination of alkaline water electrolysis with renewable energy is essential for sustainable hydrogen production without significant carbon dioxide emissions. While solar and wind energy are often favored due to their wide availability, other renewable energies, such as hydropower, biomass, and geothermal energy, are frequently utilized for the base load [7]. The direct usage of renewable energy in the power grid is difficult due to the mismatch between energy demand and production and the limited storage possibilities for electricity. Hence, excess electric energy should be chemically stored in hydrogen for later usage [6]. Due to the fluctuating and intermittent behavior of solar and wind power, alkaline water electrolyzers must be adapted to a dynamic operation. To evaluate the requirements, local weather data can be used to extract the amplitudes and frequencies of fluctuations.

Typical time-related profiles for solar radiation and wind velocity are shown in Figure 7. The data were measured by the weather station of the Clausthal University of Technology on the rooftop of a university building. Whereas the wind velocity shows a mean value of around 3.8 $m\,s^{-1}$, the significant solar radiation is only available during the daytime. Hence, the averaged value over the whole day is 233 $W\,m^{-2}$ for a sunny day and only 29 $W\,m^{-2}$ for a cloudy day.

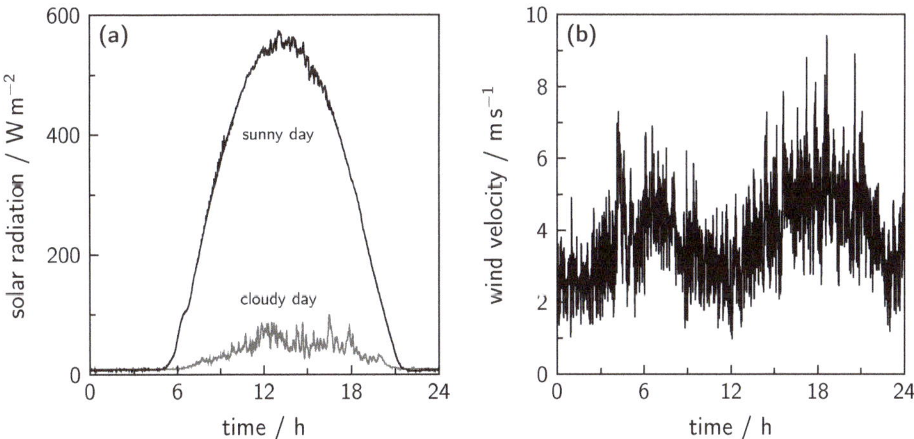

Figure 7. Typical time-related profiles for (**a**) solar radiation and (**b**) wind velocity, measured by the weather station of the Clausthal University of Technology. Though solar radiation peaks around noon, wind velocity shows sinusoidal oscillations.

The volume flow of the produced hydrogen directly follows the renewable energy profile used for operation [70]. Only a short delay is noticeable for the gas purity, which is defined by the system volume [71]. Due to the possibility of direct coupling of water electrolysis and photovoltaic panels, this technology is highly appropriate for renewable hydrogen production [29,72,73]. As photovoltaic panels require high investment costs, wind power is often favored for large-scale hydrogen production. In comparison with photovoltaic power, wind power shows a higher degree of fluctuation and is very intermittent. Therefore, the dynamic operation of alkaline water electrolyzers is more challenging [4].

Hence, the dynamic behavior of an alkaline water electrolyzer can be used to develop suitable system designs and to operate existing systems safely and efficiently. As measurements of solar radiation and wind velocity are often available for a given location, the theoretically available renewable energy can be calculated and used as an input during the system design. Different approaches exist for the calculation of solar photovoltaic power and wind turbine power. While the current–voltage characteristics of photovoltaic panels can be expressed as a function of manufacturer data and solar radiation, the power of wind turbines is a fraction of the maximum available wind power, which is defined by the wind speed and a performance coefficient [72,74].

7.1. Solar Photovoltaic Power

The behavior of photovoltaic panels can be described by single-diode and two-diode models with varying degrees of complexity. Often, the solution must be obtained iteratively or with numerical methods when very detailed models are utilized [75,76]. Simple models with analytical solutions are a recent research topic, as a short processing time can be needed for online characterization and the optimization of existing systems [75]. In Figure 8, the coupling possibilities of an alkaline water electrolyzer and solar photovoltaic panels are shown. Additional losses of a DC/DC transformer can be avoided when a direct coupling of the systems can be realized. Otherwise, the transformation ensures a fit of both systems by an indirect coupling [73,77,78].

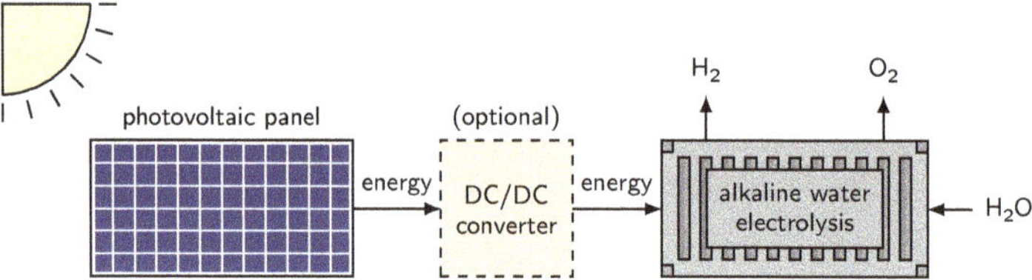

Figure 8. Schematic of alkaline water electrolysis powered by solar energy. Photovoltaic panels convert the solar radiation into electricity, which can be used for the operation. The implementation of a DC/DC power converter is optional, as direct and indirect coupling is possible [70,78,79].

When a direct coupling of both systems is to be realized, the possible operation points can be determined by the intersection of the current–voltage curves. A typical current–voltage characteristic of an alkaline water electrolyzer is given by (8). The resulting current of a photovoltaic cell I_{PV} at different solar radiation levels can be described by (10) with a suitable single-diode model as a function of the voltage U_{PV} [29,72,73]. Therefore, specific data from the photovoltaic (PV) panel and the ambient conditions are required in order to calculate the photocurrent I_{ph}, the reverse saturation current I_s, and the thermal voltage U_T. Furthermore, the serial R_s and parallel R_p resistance of the photovoltaic panel must be available.

$$I_{PV} = I_{ph} - I_s \cdot \left[\exp\left(\frac{U_{PV} + I_{PV} \cdot R_s}{U_T} \right) - 1 \right] - \frac{U_{PV} + I_{PV} \cdot R_s}{R_p} \tag{10}$$

The photocurrent I_{ph} is defined in (11), which shows a linear relationship with the solar radiation E_{sun} absorbed by the photovoltaic cell. A higher cell temperature T_c increases the photocurrent.

$$I_{ph} = \left(0.003\,\mathrm{m^2\,V^{-1}} + 10^{-7}\mathrm{m^2\,V^{-1}\,K^{-1}} \cdot T_c \right) \cdot E_{sun} \tag{11}$$

The reverse saturation current I_s can be calculated by (12) with the short-circuit current I_{sc}, the open-cell voltage U_{oc}, and the thermal voltage U_T. Whereas the short-circuit current and the open-cell voltage are provided by the manufacturer, the thermal voltage depends on the physical properties.

$$I_s = \frac{I_{sc}}{\exp\left(\frac{U_{oc}}{U_T}\right) - 1} \tag{12}$$

An equation for the thermal voltage is given in (13), which is based on the Boltzmann constant k_B ($1.3806 \cdot 10^{-23}$ J K^{-1}) and the electron charge e ($1.602\,19 \cdot 10^{-19}$ C) [72]. Additionally, the number of serially connected cells, n_s, and the cell temperature are required. Furthermore, the non-ideality factor m contains any deviations from the theoretical behavior.

$$U_T = m \cdot \frac{n_s \cdot k_B \cdot T_c}{e} \tag{13}$$

In addition to these equations, the calculation of the resulting current of a photovoltaic cell requires knowledge of the serial (R_s) and parallel (R_p) resistance of the system. By adding parallel photovoltaic cells, the current multiplies by the amount of parallel paths n_p. Suitable parameters of an existing photovoltaic cell setup are given in Table 1. For this exemplary calculation, a constant temperature of the photovoltaic cell is assumed. Otherwise, the cell temperature increases with the absorbed solar radiation. While simple linear approaches already result in a good agreement with experimental data, a complete energy balance is the best way to determine the temperature exactly [29,72].

Table 1. Parameters for the example calculation of the photovoltaic current using Equation (10). The number of serial n_s and parallel n_p connected photovoltaic cells, the short-circuit current I_{sc}, the open-cell voltage U_{oc}, the serial R_s and parallel resistance R_p, and the non-ideality factor m are setup-specific data. A constant cell temperature T_c is assumed [29,72,73].

n_s	n_p	I_{sc}	U_{oc}	R_s	R_p	m	T_c
–	–	A	V	Ω	Ω	–	°C
9	2	5.98	4.615	0.099	20	1.6	48

The results of the example calculation are shown in Figure 9. The current–voltage characteristics are given for different solar radiation levels from 200 to 1000 W m^{-2}, in combination with a typical polarization curve of an alkaline water electrolyzer (10 cm^2 electrode area) from (8) in Figure 9a. The power–voltage curves for the photovoltaic cell are shown in Figure 9b. The maximal power point (MPP) for each radiation level is marked with a dot in both diagrams.

In Figure 9a, the characteristics of the alkaline water electrolyzer deviate from the MPP curve. Therefore, the photovoltaic cell cannot deliver the maximal power, and the overall efficiency decreases. Hence, both systems should be optimized until the alkaline water electrolyzer performs close to the maximal power output [73,80]. The alternative would be an indirect coupling of both systems with the integration of a DC/DC converter, which also implies losses, with an efficiency of around 90% [81,82].

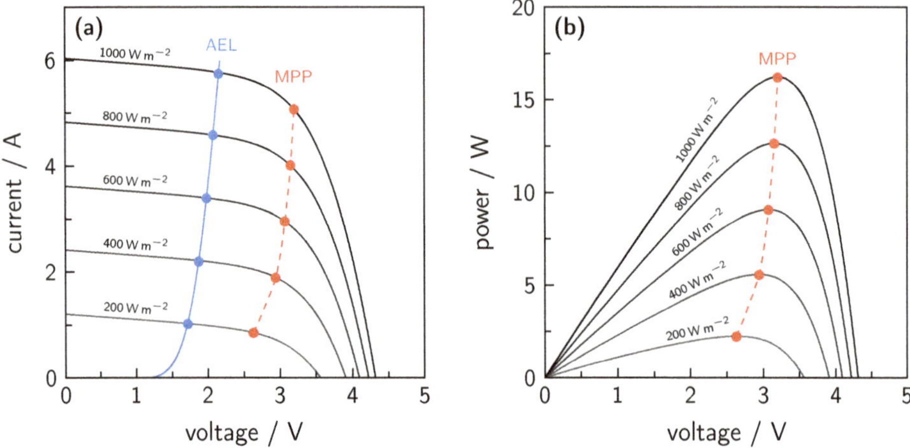

Figure 9. Example calculation results of the (**a**) current–voltage characteristics of a photovoltaic panel at different solar radiation levels and the corresponding (**b**) power–voltage curve. Additionally, a current–voltage characteristic of an alkaline water electrolyzer (AEL) is implemented. The intersections determine the possible operation points. For an efficient operation, the distance to the maximal power points (MPP) should be minimal [29,72,73].

7.2. Wind Power

As the power from photovoltaic cells is only available during the daytime, wind power is another important energy source for the renewable production of hydrogen. The schematic concept is shown in Figure 10. For the implementation of conventional wind turbines, an AC/DC converter is essential. The efficiency of an AC/DC conversion is also approximately 90% [82,83].

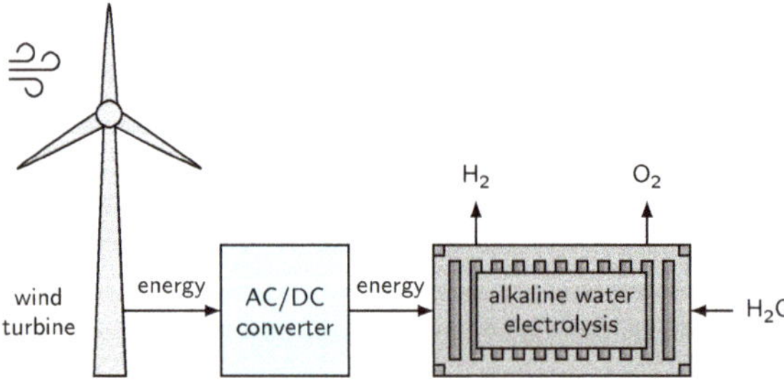

Figure 10. Schematic of alkaline water electrolysis powered by wind energy. Wind turbines convert the available wind power into electricity, which can be used for the operation. The implementation of a suitable AC/DC converter is mandatory [74,79].

For the calculation of the wind turbine power, the exact wind velocity at the height of the turbine rotor should be known. Often, the wind velocity is measured at rooftops or special measurement facilities with a defined height of approximately 10 m, which is significantly lower than the height of a wind turbine, around 100 m [84]. Therefore, the measured data should be corrected to the desired height by (14).

$$v_{\text{wind}} = v_{\text{wind,ref}} \cdot \frac{\ln\left(\frac{z_{\text{wind}}}{z_0}\right)}{\ln\left(\frac{z_{\text{wind,ref}}}{z_0}\right)} \tag{14}$$

The wind velocity v_{wind} at the height z_{wind} can be determined from the measured wind velocity $v_{\text{wind,ref}}$ at the height $z_{\text{wind,ref}}$ in combination with the roughness of the terrain z_0 [48]. To obtain the output power of a wind turbine P_{turbine}, first, the theoretical wind power P_{wind} needs to be calculated using (15). Therefore, the air density ρ (from 1.22 to 1.3 $\text{kg}\,\text{m}^{-3}$), the area spanned by the rotor blades A, and the wind velocity are needed [74,85].

$$P_{\text{wind}} = \frac{1}{2} \cdot \rho \cdot A \cdot v_{\text{wind}}^3 \tag{15}$$

The maximal wind power cannot be completely converted into wind turbine power. This circumstance is considered by the implementation of the performance coefficient C_P, which lowers the maximal reachable power output. The actual wind turbine power results from the product of the wind power and the performance coefficient in (16).

$$P_{\text{turbine}} = P_{\text{wind}} \cdot C_{\text{p}} \tag{16}$$

The determination of the correct performance coefficient is a complete research topic in itself, which consists of empirical correlations and computational fluid dynamics (CFD) simulation studies. Often, experimental data are used to fit the correlations to the measurements [74]. An example equation for the performance coefficient is shown in (17) [74,79].

$$C_{\text{p}} = 0.22 \cdot \left(\frac{116}{\lambda_{\text{i}}} - 0.4 \cdot \beta - 5\right) \cdot \exp\left(-\frac{12.5}{\lambda_{\text{i}}}\right) \tag{17}$$

Therefore, the pitch angle of the turbine blades β has to be defined and the tip speed ratio λ needs to be calculated in (18) from the turbine blade radius R, the rotational speed ω, and the wind speed [74].

$$\lambda = \frac{R \cdot \omega}{v_{\text{wind}}} \tag{18}$$

The calculation of the performance coefficient also requires the parameter λ_{i}, which is described by (19) based on the tip speed ratio and the blade pitch angle [74].

$$\frac{1}{\lambda_{\text{i}}} = \frac{1}{\lambda + 0.08 \cdot \beta} - \frac{0.035}{\beta^3 + 1} \tag{19}$$

For the blade radius, a value of 46.5 m is assumed, which is a typical blade length for a wind turbine with a rated power of 2 MW [74]. In Figure 11, the calculation results for the performance coefficient, depending on the tip speed ratio and the turbine power at different wind velocities, are shown. The performance coefficient of conventional wind turbines is limited at $C_{\text{p}} = 0.593$ [74]. In this example, a maximal performance coefficient of approximately $C_{\text{p}} = 0.450$ is reached for a blade pitch angle of $\beta = 0°$. With an increasing pitch angle, the maximum of the performance coefficient decreases and shifts towards smaller tip speed ratios.

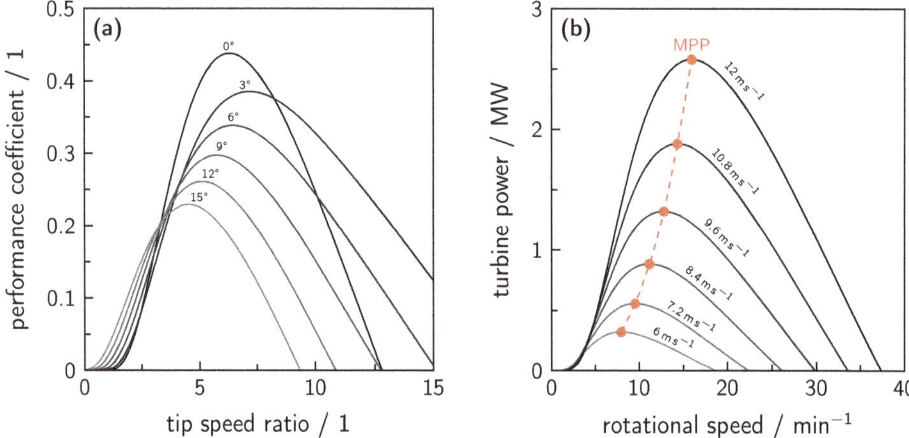

Figure 11. Example calculation results of (**a**) the performance coefficient for various rotor blade pitch angles using Equation (17) and (**b**) the wind turbine power for different wind velocities using Equation (16). The maximum power point (MPP) trajectory is marked [74,79].

For the calculation of the turbine power in Figure 11b, a pitch angle of $\beta = 6°$ is assumed. With increasing wind velocity, the value of the maximal power point (MPP) becomes higher and shifts towards faster rotational speeds. The rated wind speed of this exemplary wind turbine is at $11\,\mathrm{m\,s^{-1}}$ with rotational speeds from 6 to $17\,\mathrm{min^{-1}}$. The cut-in wind speed is $3\,\mathrm{m\,s^{-1}}$ and the cut-out wind speed is $22\,\mathrm{m\,s^{-1}}$ [74]. In comparison with the power characteristics of photovoltaic panels, the polarization curve of alkaline water electrolyzers can not be directly optimized towards the MPP trajectory, as the optimal operation point highly depends on the wind turbine design and weather conditions. Therefore, an efficient AC/DC converter is the best option for maintaining an efficient operation of an alkaline water electrolyzer [82].

8. Hydrogen Energy System and Power Grid Stabilization

An exemplary process scheme for a hydrogen energy system is provided in Figure 12. Photovoltaic panels and wind turbines are connected with suitable converters to a DC bus, from which alkaline water electrolyzers are powered. The produced hydrogen can be stored for later application in fuel cells. To raise the fuel cell efficiency, the produced oxygen can be used instead of air. Therefore, an additional storage tank must be available, which incurs further costs [86].

The fuel cells are also connected to the DC bus, and the power can be used by the electricity grid with DC/AC converters. At lower energy demands, hydrogen can be produced and converted back into energy when it is needed. As conventional alkaline water electrolyzers are designed for operation at constant conditions, occurring fluctuations may be damped by additional energy storage devices like batteries, supercapacitors, or flywheels [25,28,82]. When excess energy is available, this energy storage can be charged to be fully available when needed. The damping quantity is limited to a certain degree of fluctuation, as the energy storage amount is also restricted to the capacity of all installed devices. Additionally, the produced hydrogen can also be used for the decarbonization of industrial processes or as a fuel in the transport sector [87–89]. To raise the overall efficiency, some DC/DC converters could be neglected by optimized system designs by lowering the system flexibility. Furthermore, when the alkaline water electrolyzers are able to operate under dynamic conditions, additional energy storage devices are not required or, at least, the number of such devices could be lowered. There are still some challenges for electrolyzer manufacturers to overcome before this possibility becomes available.

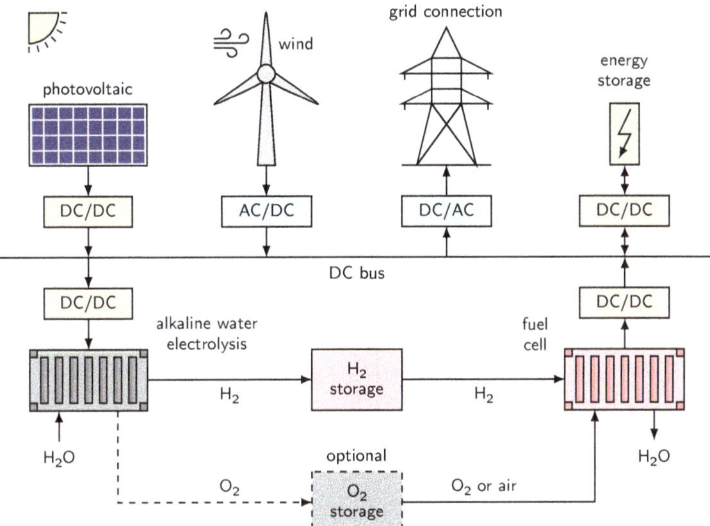

Figure 12. The schematic process scheme of a hydrogen energy system. Photovoltaic panels and wind turbines generate renewable energy to power alkaline water electrolyzers, and stored hydrogen can be converted back into electricity by fuel cells. Therefore, either oxygen or air can be utilized. Additional energy storage devices can damp fluctuations, and the complete hydrogen energy system can be used for power grid stabilization [25,28,82,87].

With an increasing share of renewable energies in the power grid, it is difficult to maintain a constant power frequency. Such hydrogen energy systems or alkaline water electrolyzers can be used to stabilize the power frequency by damping the fluctuations. An additional benefit would be the reduction of the conventional spinning reserve, which reduces costs and CO_2 emissions [87,90]. A predictive control can be used for stable and efficient operation. Pressurized alkaline electrolyzers are more suitable for damping fast fluctuations, whereas atmospheric units can handle the slow fluctuations [87].

9. Limitations and Solution Approaches

The implementation of a hydrogen energy system into the existing power grid is a challenging task with some limitations which must be overcome in order to guarantee high system availability. The main problem of an alkaline water electrolyzer powered by renewable energy is the high gas impurity in the part-load range, which can cause a safety shutdown when reaching a foreign gas contamination of 2 vol.% [31,91]. Hence, the annual operation time is limited to the time spans with sufficient renewable energy [91].

9.1. Limited Operation Time

The limited operation time leads to a high number of startup and shutdown cycles, which can exceed the maximal start/stop count defined by the manufacturer and, therefore, can lower the expected system lifetime or warranty agreements. Mainly, the electrodes are affected by the repetitive start/stop behavior and the electrode degradation is accelerated [48,82]. Nickel electrodes are known to degrade significantly after 5000 to 10,000 start/stop cycles. When operating with photovoltaic power only, 7000 to 11,000 cycles are already reached in the period of 20 to 30 years. The fluctuating nature of renewable energy amplifies the electrode degradation, as this phenomenon acts partly as a start/stop process [92]. This issue can be solved by the development of stable electrode compositions or self-repairing electrode surfaces [92].

To circumvent the drawbacks of having only one renewable power source, such as in the daytime-limited operation with solar power, the combination of several energy sources enhances the overall efficiency. While the operation with only PV shows a faradaic efficiency of approximately 40%, wind power leads to a faradaic efficiency of around 80%. The combination of both technologies enhances the faradaic efficiency above 85% [79].

To hinder the gas impurity from reaching the lower explosion limit, the part-load range of most alkaline electrolyzers is limited to 10 to 25% of their nominal load [82,91]. Fluctuations below the minimal load can be balanced out with the implementation of energy storage devices, as shown in Figure 12; however, in some scenarios, the available energy storage will not be sufficient. When the gas impurity is still in a tolerable region, short periods without an electrode polarization can be allowed. The cathode starts to degrade noticeably below a voltage of around 0.25 V [82]. Thus, the complete shutdown can be held until reaching this voltage limit. The available time depends on the electrode composition, as the electrochemical double layer acts as a capacitor and delays the voltage breakdown after a power loss. Experimentally, a time span of around 10 min has been reported [82].

9.2. Optimal System Design and Operation Strategies

To mitigate the rise of gas impurities during low power availability, an optimal system design can allow enough time until sufficient energy is available again. While the gas volume inside the system acts as a buffer tank and dilutes the gas contamination, the liquid and the solid volume of an electrolyzer buffers the system temperature during part-load operation [25,71].

Furthermore, to maintain an efficient operation, the system temperature has to be in an optimal range of 50 to 80 °C for an electrolyte solution with 20 to 30 wt.% KOH [25]. As most renewable-energy-powered alkaline water electrolyzers will not provide a separate heating unit, the temperature needs to be reached and maintained only by the heat of the reaction [4]. Temperatures above 80 °C should be avoided with a suitable cooling system to prevent high degradation rates. An alternative would be the operation at low temperatures to damp electrode degradation, but then, very active electrocatalysts are needed to reach a sufficient efficiency [86].

More experimental and theoretical work is needed to fully understand the dynamic behavior of alkaline water electrolyzers powered by renewable energy [25]. In addition to an optimal system design, suitable dynamic operation strategies can be beneficial for lowering the gas impurity. While low gas impurities occur with separated electrolyte cycles, high gas impurities result in combined mode. The measured stationary gas impurities in Figure 6 are reached after a specific duration. When the electrolyzer is able to switch between both operation modes automatically, this can be used to switch to the separated mode when the gas impurity is too high, and then combine again when a sufficient gas production rate is available. Experimental work shows that the gas impurity can be almost halved by this approach [31].

The primary reason for high gas contamination is the continuous operation at low current densities. This circumstance can be prevented by reducing the overall cell area (overloading) or by subdividing the system into several smaller blocks [91]. While the implementation of electrolyzers with smaller electrode areas also limits the maximal load compared to larger systems, partial system operation is a more elegant method. During low power availability, single stacks or compartments of a system with multiple stacks can be powered off, which lowers the available electrode area and therefore results in higher current densities [93]. Obviously, this strategy causes problems in maintaining the optimal system temperature due to the disabled components. An alternative method to prevent adverse process states is the use of predictive control systems. For example, when low renewable power availability is forecasted, the system can change the temperature, pressure, or operation mode to a more suitable state before negative effects occur [87].

10. Conclusions

The combination of alkaline water electrolysis and renewable energy for sustainable hydrogen production is an essential step towards the decarbonization of industrial processes and the transport sector [87–89]. To determine the most relevant limitations and to propose suitable solution approaches, the technologies have to be fully understood [25]. Whereas the process of alkaline water electrolysis can be defined by current–voltage characteristics and the resulting gas impurity, photovoltaic panels and wind turbines should be operated at the maximal power point [73,74,79]. Therefore, the influencing parameters must be known. Different model approaches exist, out of which the most suitable one should be chosen. While empirical correlations are often only valid for the specific experimental setup, physically reasonable models can be used in a more general way to develop new solutions. For alkaline water electrolysis, many experimental and theoretical data are available to calculate and analyze the cell voltage under operation conditions. As the actual system design and cell arrangement differ for every electrolyzer, certain parameters have to be determined experimentally to use the proposed models for another system. Mainly, this issue exists for electrode compositions and separator materials. To describe the gas purity of hydrogen and oxygen mathematically, only models and correlations on an empirical basis are currently available due to the high number of influencing variables [31,32]. As the gas impurity mainly determines the system availability of an alkaline water electrolyzer, more research for the development of physically-based models is needed. The dynamic system behavior should be analyzed, as optimized dynamic operation strategies can be beneficial for the overall system efficiency. Many models with different complexity levels are available for the description of the current–voltage characteristics of photovoltaic panels. Most models rely on physical principles and manufacturer data [75]. Thus, proper modeling for different systems is possible. The power conversion by wind turbines can be described by system properties and suitable correlations for the performance coefficient [74]. As this variable is influenced by many parameters, including the design of the turbine blades, the correlation should only be used for very similar wind turbines, or the parameters must be determined experimentally or by simulation.

To conclude, there are appropriate models available for all components of a hydrogen energy system. However, some descriptions need further improvement to be applicable to a variety of different system designs. With this knowledge and with experimental studies, many researchers have already examined the limitations of renewable-powered alkaline water electrolyzers [48,79,82]. The central prospect is to increase the operation time through intelligent system designs and advantageous operational concepts. While the implementation of conventional energy storage devices to damp the dynamics is a first logical step, alkaline water electrolyzers should be enabled to handle all dynamics directly to reduce costs and to enhance the efficiency [25]. As the hydrogen production from fossil energy carriers is less expensive than hydrogen from electrolysis processes, only optimized systems with the use of excess renewable energy can be competitive.

Author Contributions: Conceptualization, methodology, software, validation, formal analysis, investigation, data curation, writing—original draft preparation, visualization, J.B.; resources, writing—review and editing, J.B. and T.T.; supervision, project administration, funding acquisition, T.T. All authors have read and agreed to the published version of the manuscript.

Funding: This work is funded by the Deutsche Forschungsgemeinschaft (DFG, German Research Foundation) Project numbers: 290019031; 391348959.

Acknowledgments: The authors thank the Institute of Electrical Information Technology (IEI) of the Clausthal University of Technology for providing the weather data.

Conflicts of Interest: The authors declare no conflict of interest. The funders had no role in the design of the study; in the collection, analyses, or interpretation of data; in the writing of the manuscript, or in the decision to publish the results.

Abbreviations

AC	Alternating current
AEL	Alkaline water electrolysis
CFD	Computational fluid dynamics
DC	Direct current
HTEL	High-temperature electrolysis
ILs	Ionic liquids
MPP	Maximum power point
PEMEL	Proton exchange membrane electrolysis
PV	Photovoltaic
SOEL	Solid oxide electrolysis

Appendix A. Correlations and Parameters

A correlation for the reversible cell voltage U_{rev} of alkaline water electrolysis is given in (A1). The obtained value can be used for the calculation of the cell voltage in (7) or (8) at atmospheric conditions. For a pressurized system, extended correlations are required, as the reversible cell voltage increases at higher pressures [40]. The empirical correlation parameters for the calculation of cell voltage by (7) and (8) are given in Table A1.

$$U_{rev} = 1.503\,42\,\mathrm{V} - 9.956 \cdot 10^{-4}\,\mathrm{V} \cdot \left(\frac{T}{K}\right) + 2.5 \cdot 10^{-7}\,\mathrm{V} \cdot \left(\frac{T}{K}\right)^2 \tag{A1}$$

Table A1. Parameters for the calculation of cell voltage by Equations (7) and (8) [28,39,94].

Parameter	Equation (7) [28,94]	Equation (8) [39]	Unit
r_1	$8.05 \cdot 10^{-5}$	$4.451\,53 \cdot 10^{-5}$	$\Omega\,m^2$
r_2	$-2.5 \cdot 10^{-7}$	$6.888\,74 \cdot 10^{-9}$	$\Omega\,m^2\,°C^{-1}$
s	0.185	0.338 24	V
t_1	1.002	−0.015 39	$m^2\,A^{-1}$
t_2	8.424	2.001 81	$m^2\,°C\,A^{-1}$
t_3	247.3	15.241 78	$m^2\,°C^2\,A^{-1}$
d_1	–	$-3.129\,96 \cdot 10^{-6}$	$\Omega\,m^2$
d_2	–	$4.471\,37 \cdot 10^{-7}$	$\Omega\,m^2\,bar^{-1}$

The correlations for the calculation of specific electrolyte conductivity for KOH and NaOH can be found in (A2) and (A3). The required correlation parameters are listed in Table A2. The validity range for (A2) is a temperature T from 258.15 to 373.15 K and KOH mass fractions w_{KOH} between 0.15 and 0.45. Equation (A3) is valid for temperatures ϑ between 25 and 50 °C and NaOH mass fractions w_{NaOH} from 0.08 to 0.25 [52,57].

$$\begin{aligned}\sigma_{KOH} =& K_1 \cdot (100 \cdot w_{KOH}) + K_2 \cdot T + K_3 \cdot T^2 + K_4 \cdot T \cdot (100 \cdot w_{KOH}) \\ &+ K_5 \cdot T^2 \cdot (100 \cdot w_{KOH})^{K_6} + K_7 \cdot \frac{T}{(100 \cdot w_{KOH})} + K_8 \cdot \frac{(100 \cdot w_{KOH})}{T}\end{aligned} \tag{A2}$$

$$\sigma_{NaOH} = K_1 + K_2 \cdot \vartheta + K_3 \cdot w_{NaOH}^3 + K_4 \cdot w_{NaOH}^2 + K_5 \cdot w_{NaOH} \tag{A3}$$

Table A2. Parameters for the calculation of the specific electrolyte conductivities of KOH and NaOH solutions by Equations (A2) and (A3) [52,57].

Parameter	Equation (A2) [57]	Unit	Equation (A3) [52]	Unit
K_1	27.984 480 3	$S\,m^{-1}$	−45.7	$S\,m^{-1}$
K_2	−0.924 129 482	$S\,m^{-1}\,K^{-1}$	1.02	$S\,m^{-1}\,°C^{-1}$
K_3	−0.014 966 037 1	$S\,m^{-1}\,K^{-2}$	3200	$S\,m^{-1}$
K_4	−0.090 520 955 1	$S\,m^{-1}\,K^{-1}$	−2990	$S\,m^{-1}$
K_5	0.011 493 325 2	$S\,m^{-1}\,K^{-2}$	784	$S\,m^{-1}$
K_6	0.1765	-	–	–
K_7	6.966 485 18	$S\,m^{-1}\,K^{-1}$	–	–
K_8	−2898.156 58	$S\,K\,m^{-1}$	–	–

References

1. Sherif, S.; Barbir, F.; Veziroglu, T. Towards a Hydrogen Economy. *Electr. J.* **2005**, *18*, 62–76. [CrossRef]
2. Suleman, F.; Dincer, I.; Agelin-Chaab, M. Environmental Impact Assessment and Comparison of Some Hydrogen Production Options. *Int. J. Hydrog. Energy* **2015**, *40*, 6976–6987. [CrossRef]
3. Dincer, I.; Acar, C. Review and Evaluation of Hydrogen Production Methods for Better Sustainability. *Int. J. Hydrog. Energy* **2015**, *40*, 11094–11111. [CrossRef]
4. Gandía, L.M.; Oroz, R.; Ursúa, A.; Sanchis, P.; Diéguez, P.M. Renewable Hydrogen Production: Performance of an Alkaline Water Electrolyzer Working under Emulated Wind Conditions. *Energy Fuels* **2007**, *21*, 1699–1706. [CrossRef]
5. Santos, D.M.F.; Sequeira, C.A.C.; Figueiredo, J.L. Hydrogen Production by Alkaline Water Electrolysis. *Quim. Nova* **2013**, *36*, 1176–1193. [CrossRef]
6. Wang, M.; Wang, Z.; Gong, X.; Guo, Z. The Intensification Technologies to Water Electrolysis for Hydrogen Production—A Review. *Renew. Sustain. Energy Rev.* **2014**, *29*, 573–588. [CrossRef]
7. Gahleitner, G. Hydrogen from Renewable Electricity: An International Review of Power-to-Gas Pilot Plants for Stationary Applications. *Int. J. Hydrog. Energy* **2013**, *38*, 2039–2061. [CrossRef]
8. Carpetis, C. Estimation of Storage Costs for Large Hydrogen Storage Facilities. *Int. J. Hydrog. Energy* **1982**, *7*, 191–203. [CrossRef]
9. Michalski, J.; Bünger, U.; Crotogino, F.; Donadei, S.; Schneider, G.S.; Pregger, T.; Cao, K.K.; Heide, D. Hydrogen Generation by Electrolysis and Storage in Salt Caverns: Potentials, Economics and Systems Aspects with Regard to the German Energy Transition. *Int. J. Hydrog. Energy* **2017**, *42*, 13427–13443. [CrossRef]
10. Pellow, M.A.; Emmott, C.J.M.; Barnhart, C.J.; Benson, S.M. Hydrogen or Batteries for Grid Storage? A Net Energy Analysis. *Energy Environ. Sci.* **2015**, *8*, 1938–1952. [CrossRef]
11. Qadrdan, M.; Abeysekera, M.; Chaudry, M.; Wu, J.; Jenkins, N. Role of Power-to-Gas in an Integrated Gas and Electricity System in Great Britain. *Int. J. Hydrog. Energy* **2015**, *40*, 5763–5775. [CrossRef]
12. Schouten, J.; Janssenvanrosmalen, R.; Michels, J. Modeling Hydrogen Production for Injection into the Natural Gas Grid: Balance between Production, Demand and Storage. *Int. J. Hydrog. Energy* **2006**, *31*, 1698–1706. [CrossRef]
13. Goncharov, A.; Guglya, A.; Melnikova, E. On the Feasibility of Developing Hydrogen Storages Capable of Adsorption Hydrogen Both in Its Molecular and Atomic States. *Int. J. Hydrog. Energy* **2012**, *37*, 18061–18073. [CrossRef]
14. Goncharov, A.; Guglya, A.; Melnikova, E. Corrigendum to "On the Feasibility of Developing Hydrogen Storages Capable of Adsorption Hydrogen Both in Its Molecular and Atomic States" [Int J Hydrogen Energy, 37 (2012) 18061–18073]. *Int. J. Hydrog. Energy* **2013**, *38*, 3521. [CrossRef]
15. Schmidt, O.; Gambhir, A.; Staffell, I.; Hawkes, A.; Nelson, J.; Few, S. Future Cost and Performance of Water Electrolysis: An Expert Elicitation Study. *Int. J. Hydrog. Energy* **2017**, *42*, 30470–30492. [CrossRef]
16. Schalenbach, M.; Zeradjanin, A.R.; Kasian, O.; Cherevko, S.; Mayrhofer, K.J. A Perspective on Low-Temperature Water Electrolysis-Challenges in Alkaline and Acidic Technology. *Int. J. Electrochem. Sci.* **2018**, *13*, 1173–1226. [CrossRef]

17. Zeng, K.; Zhang, D. Recent Progress in Alkaline Water Electrolysis for Hydrogen Production and Applications. *Prog. Energy Combust. Sci.* **2010**, *36*, 307–326. [CrossRef]
18. Carmo, M.; Fritz, D.L.; Mergel, J.; Stolten, D. A Comprehensive Review on PEM Water Electrolysis. *Int. J. Hydrog. Energy* **2013**, *38*, 4901–4934. [CrossRef]
19. David, M.; Ocampo-Martínez, C.; Sánchez-Peña, R. Advances in Alkaline Water Electrolyzers: A Review. *J. Energy Storage* **2019**, *23*, 392–403. [CrossRef]
20. Buttler, A.; Spliethoff, H. Current Status of Water Electrolysis for Energy Storage, Grid Balancing and Sector Coupling via Power-to-Gas and Power-to-Liquids: A Review. *Renew. Sustain. Energy Rev.* **2018**, *82*, 2440–2454. [CrossRef]
21. Rashid, M.; Mesfer, M.K.A.; Naseem, H.; Danish, M. Hydrogen Production by Water Electrolysis: A Review of Alkaline Water Electrolysis, PEM Water Electrolysis and High Temperature Water Electrolysis. *Int. J. Eng. Adv. Technol.* **2015**, *4*, 80–93.
22. Götz, M.; Lefebvre, J.; Mörs, F.; McDaniel Koch, A.; Graf, F.; Bajohr, S.; Reimert, R.; Kolb, T. Renewable Power-to-Gas: A Technological and Economic Review. *Renew. Energy* **2016**, *85*, 1371–1390. [CrossRef]
23. Marini, S.; Salvi, P.; Nelli, P.; Pesenti, R.; Villa, M.; Berrettoni, M.; Zangari, G.; Kiros, Y. Advanced Alkaline Water Electrolysis. *Electrochim. Acta* **2012**, *82*, 384–391. [CrossRef]
24. Seibel, C.; Kuhlmann, J.W. Dynamic Water Electrolysis in Cross-Sectoral Processes. *Chem. Ing. Tech.* **2018**, *90*, 1430–1436. [CrossRef]
25. Shen, X.; Zhang, X.; Li, G.; Lie, T.T.; Hong, L. Experimental Study on the External Electrical Thermal and Dynamic Power Characteristics of Alkaline Water Electrolyzer. *Int. J. Energy Res.* **2018**, *42*, 3244–3257. [CrossRef]
26. Clarivate Analytics. Web of Science Database. 2020. Available online: http://apps.webofknowledge.com (accessed on 14 January 2020).
27. Ehret, O.; Bonhoff, K. Hydrogen as a Fuel and Energy Storage: Success Factors for the German Energiewende. *Int. J. Hydrog. Energy* **2015**, *40*, 5526–5533. [CrossRef]
28. Ulleberg, O. Modeling of Advanced Alkaline Electrolyzers: A System Simulation Approach. *Int. J. Hydrog. Energy* **2003**, *28*, 21–33. [CrossRef]
29. Đukić, A.; Firak, M. Hydrogen Production Using Alkaline Electrolyzer and Photovoltaic (PV) Module. *Int. J. Hydrog. Energy* **2011**, *36*, 7799–7806. [CrossRef]
30. LeRoy, R.L. The Thermodynamics of Aqueous Water Electrolysis. *J. Electrochem. Soc.* **1980**, *127*, 1954. [CrossRef]
31. Haug, P.; Koj, M.; Turek, T. Influence of Process Conditions on Gas Purity in Alkaline Water Electrolysis. *Int. J. Hydrog. Energy* **2017**, *42*, 9406–9418. [CrossRef]
32. Haug, P.; Kreitz, B.; Koj, M.; Turek, T. Process Modelling of an Alkaline Water Electrolyzer. *Int. J. Hydrog. Energy* **2017**, *42*, 15689–15707. [CrossRef]
33. Renaud, R.; Leroy, R. Separator Materials for Use in Alkaline Water Electrolysers. *Int. J. Hydrog. Energy* **1982**, *7*, 155–166. [CrossRef]
34. Kraglund, M.R.; Aili, D.; Jankova, K.; Christensen, E.; Li, Q.; Jensen, J.O. Zero-Gap Alkaline Water Electrolysis Using Ion-Solvating Polymer Electrolyte Membranes at Reduced KOH Concentrations. *J. Electrochem. Soc.* **2016**, *163*, F3125–F3131. [CrossRef]
35. Kraglund, M.R.; Carmo, M.; Schiller, G.; Ansar, S.A.; Aili, D.; Christensen, E.; Jensen, J.O. Ion-Solvating Membranes as a New Approach towards High Rate Alkaline Electrolyzers. *Energy Environ. Sci.* **2019**, *12*, 3313–3318. [CrossRef]
36. Hnát, J.; Paidar, M.; Schauer, J.; Žitka, J.; Bouzek, K. Polymer Anion-Selective Membranes for Electrolytic Splitting of Water. Part II: Enhancement of Ionic Conductivity and Performance under Conditions of Alkaline Water Electrolysis. *J. Appl. Electrochem.* **2012**, *42*, 545–554. [CrossRef]
37. Hnát, J.; Plevová, M.; Žitka, J.; Paidar, M.; Bouzek, K. Anion-Selective Materials with 1,4-Diazabicyclo[2.2.2]Octane Functional Groups for Advanced Alkaline Water Electrolysis. *Electrochim. Acta* **2017**, *248*, 547–555. [CrossRef]
38. Phillips, R.; Dunnill, C.W. Zero Gap Alkaline Electrolysis Cell Design for Renewable Energy Storage as Hydrogen Gas. *RSC Adv.* **2016**, *6*, 100643–100651. [CrossRef]

39. Sánchez, M.; Amores, E.; Rodríguez, L.; Clemente-Jul, C. Semi-Empirical Model and Experimental Validation for the Performance Evaluation of a 15 kW Alkaline Water Electrolyzer. *Int. J. Hydrog. Energy* **2018**, *43*, 20332–20345. [CrossRef]
40. Hammoudi, M.; Henao, C.; Agbossou, K.; Dubé, Y.; Doumbia, M. New Multi-Physics Approach for Modelling and Design of Alkaline Electrolyzers. *Int. J. Hydrog. Energy* **2012**, *37*, 13895–13913. [CrossRef]
41. Koj, M.; Gimpel, T.; Schade, W.; Turek, T. Laser Structured Nickel-Iron Electrodes for Oxygen Evolution in Alkaline Water Electrolysis. *Int. J. Hydrog. Energy* **2019**, *44*, 12671–12684. [CrossRef]
42. Koj, M.; Qian, J.; Turek, T. Novel Alkaline Water Electrolysis with Nickel-Iron Gas Diffusion Electrode for Oxygen Evolution. *Int. J. Hydrog. Energy* **2019**, *44*, 29862–29875. [CrossRef]
43. Hall, D.E. Electrodes for Alkaline Water Electrolysis. *J. Electrochem. Soc.* **1981**, *128*, 740. [CrossRef]
44. Huot, J.Y. Low Hydrogen Overpotential Nanocrystalline Ni-Mo Cathodes for Alkaline Water Electrolysis. *J. Electrochem. Soc.* **1991**, *138*, 1316. [CrossRef]
45. Rauscher, T.; Bernäcker, C.I.; Mühle, U.; Kieback, B.; Röntzsch, L. The Effect of Fe as Constituent in Ni-Base Alloys on the Oxygen Evolution Reaction in Alkaline Solutions at High Current Densities. *Int. J. Hydrog. Energy* **2019**, *44*, 6392–6402. [CrossRef]
46. Fan, C. Study of Electrodeposited Nickel-Molybdenum, Nickel-Tungsten, Cobalt-Molybdenum, and Cobalt-Tungsten as Hydrogen Electrodes in Alkaline Water Electrolysis. *J. Electrochem. Soc.* **1994**, *141*, 382. [CrossRef]
47. Rauscher, T.; Müller, C.I.; Schmidt, A.; Kieback, B.; Röntzsch, L. Ni–Mo–B Alloys as Cathode Material for Alkaline Water Electrolysis. *Int. J. Hydrog. Energy* **2016**, *41*, 2165–2176. [CrossRef]
48. Ursúa, A.; San Martín, I.; Barrios, E.L.; Sanchis, P. Stand-Alone Operation of an Alkaline Water Electrolyser Fed by Wind and Photovoltaic Systems. *Int. J. Hydrog. Energy* **2013**, *38*, 14952–14967. [CrossRef]
49. Henao, C.; Agbossou, K.; Hammoudi, M.; Dubé, Y.; Cardenas, A. Simulation Tool Based on a Physics Model and an Electrical Analogy for an Alkaline Electrolyser. *J. Power Sources* **2014**, *250*, 58–67. [CrossRef]
50. Balabel, A.; Zaky, M.S.; Sakr, I. Optimum Operating Conditions for Alkaline Water Electrolysis Coupled with Solar PV Energy System. *Arab. J. Sci. Eng.* **2014**, *39*, 4211–4220. [CrossRef]
51. Vermeiren, P. Zirfon®: A New Separator for Ni-H_2 Batteries and Alkaline Fuel Cells. *Int. J. Hydrog. Energy* **1996**, *21*, 679–684. [CrossRef]
52. Le Bideau, D.; Mandin, P.; Benbouzid, M.; Kim, M.; Sellier, M. Review of Necessary Thermophysical Properties and Their Sensivities with Temperature and Electrolyte Mass Fractions for Alkaline Water Electrolysis Multiphysics Modelling. *Int. J. Hydrog. Energy* **2019**, *44*, 4553–4569. [CrossRef]
53. Shoor, S.K.; Walker, R.D.; Gubbins, K.E. Salting out of Nonpolar Gases in Aqueous Potassium Hydroxide Solutions. *J. Phys. Chem.* **1969**, *73*, 312–317. [CrossRef]
54. Grover, P.K.; Ryall, R.L. Critical Appraisal of Salting-Out and Its Implications for Chemical and Biological Sciences. *Chem. Rev.* **2005**, *105*, 1–10. [CrossRef] [PubMed]
55. Randall, M.; Failey, C.F. The Activity Coefficient of Non-Electrolytes in Aqueous Salt Solutions from Solubility Measurements. The Salting-out Order of the Ions. *Chem. Rev.* **1927**, *4*, 285–290. [CrossRef]
56. Lang, W.; Zander, R. Salting-out of Oxygen from Aqueous Electrolyte Solutions: Prediction and Measurement. *Ind. Eng. Chem. Fundam.* **1986**, *25*, 775–782. [CrossRef]
57. See, D.M.; White, R.E. Temperature and Concentration Dependence of the Specific Conductivity of Concentrated Solutions of Potassium Hydroxide. *J. Chem. Eng. Data* **1997**, *42*, 1266–1268. [CrossRef]
58. de Souza, R.F.; Padilha, J.C.; Gonçalves, R.S.; Rault-Berthelot, J. Dialkylimidazolium Ionic Liquids as Electrolytes for Hydrogen Production from Water Electrolysis. *Electrochem. Commun.* **2006**, *8*, 211–216. [CrossRef]
59. Zhao, Y.; Zhao, J.; Huang, Y.; Zhou, Q.; Zhang, X.; Zhang, S. Toxicity of Ionic Liquids: Database and Prediction via Quantitative Structure–Activity Relationship Method. *J. Hazard. Mater.* **2014**, *278*, 320–329. [CrossRef]
60. Zhao, Y.; Gani, R.; Afzal, R.M.; Zhang, X.; Zhang, S. Ionic Liquids for Absorption and Separation of Gases: An Extensive Database and a Systematic Screening Method. *AIChE J.* **2017**, *63*, 1353–1367. [CrossRef]
61. Zhao, Y.; Pan, M.; Kang, X.; Tu, W.; Gao, H.; Zhang, X. Gas Separation by Ionic Liquids: A Theoretical Study. *Chem. Eng. Sci.* **2018**, *189*, 43–55. [CrossRef]

62. De Souza, R.F.; Padilha, J.C.; Gonçalves, R.S.; de Souza, M.O.; Rault-Berthelot, J. Electrochemical Hydrogen Production from Water Electrolysis Using Ionic Liquid as Electrolytes: Towards the Best Device. *J. Power Sources* **2007**, *164*, 792–798. [CrossRef]
63. Schalenbach, M.; Lueke, W.; Stolten, D. Hydrogen Diffusivity and Electrolyte Permeability of the Zirfon PERL Separator for Alkaline Water Electrolysis. *J. Electrochem. Soc.* **2016**, *163*, F1480–F1488. [CrossRef]
64. Trinke, P.; Haug, P.; Brauns, J.; Bensmann, B.; Hanke-Rauschenbach, R.; Turek, T. Hydrogen Crossover in PEM and Alkaline Water Electrolysis: Mechanisms, Direct Comparison and Mitigation Strategies. *J. Electrochem. Soc.* **2018**, *165*, F502–F513. [CrossRef]
65. Roy, A.; Watson, S.; Infield, D. Comparison of Electrical Energy Efficiency of Atmospheric and High-Pressure Electrolysers. *Int. J. Hydrog. Energy* **2006**, *31*, 1964–1979. [CrossRef]
66. Ursúa, A.; Sanchis, P. Static–Dynamic Modelling of the Electrical Behaviour of a Commercial Advanced Alkaline Water Electrolyser. *Int. J. Hydrog. Energy* **2012**, *37*, 18598–18614. [CrossRef]
67. Dobó, Z.; Palotás, Á.B. Impact of the Voltage Fluctuation of the Power Supply on the Efficiency of Alkaline Water Electrolysis. *Int. J. Hydrog. Energy* **2016**, *41*, 11849–11856. [CrossRef]
68. Dobó, Z.; Palotás, Á.B. Impact of the Current Fluctuation on the Efficiency of Alkaline Water Electrolysis. *Int. J. Hydrog. Energy* **2017**, *42*, 5649–5656. [CrossRef]
69. Speckmann, F.W.; Bintz, S.; Birke, K.P. Influence of Rectifiers on the Energy Demand and Gas Quality of Alkaline Electrolysis Systems in Dynamic Operation. *Appl. Energy* **2019**, *250*, 855–863. [CrossRef]
70. De Fátima Palhares, D.D.; Vieira, L.G.M.; Damasceno, J.J.R. Hydrogen Production by a Low-Cost Electrolyzer Developed through the Combination of Alkaline Water Electrolysis and Solar Energy Use. *Int. J. Hydrog. Energy* **2018**, *43*, 4265–4275. [CrossRef]
71. Hug, W.; Bussmann, H.; Brinner, A. Intermittent Operation and Operation Modeling of an Alkaline Electrolyzer. *Int. J. Hydrog. Energy* **1993**, *18*, 973–977. [CrossRef]
72. Firak, M.; Djukic, A. An Investigation into the Effect of Photovoltaic Module Electric Properties on Maximum Power Point Trajectory with the Aim of Its Alignment with Electrolyzer U-I Characteristic. *Therm. Sci.* **2010**, *14*, 729–738. [CrossRef]
73. Đukić, A. Autonomous Hydrogen Production System. *Int. J. Hydrog. Energy* **2015**, *40*, 7465–7474. [CrossRef]
74. Dai, J.; Liu, D.; Wen, L.; Long, X. Research on Power Coefficient of Wind Turbines Based on SCADA Data. *Renew. Energy* **2016**, *86*, 206–215. [CrossRef]
75. Chin, V.J.; Salam, Z.; Ishaque, K. Cell Modelling and Model Parameters Estimation Techniques for Photovoltaic Simulator Application: A Review. *Appl. Energy* **2015**, *154*, 500–519. [CrossRef]
76. Vergura, S. A Complete and Simplified Datasheet-Based Model of PV Cells in Variable Environmental Conditions for Circuit Simulation. *Energies* **2016**, *9*, 326. [CrossRef]
77. Kovač, A.; Marciuš, D.; Budin, L. Solar Hydrogen Production via Alkaline Water Electrolysis. *Int. J. Hydrog. Energy* **2019**, *44*, 9841–9848. [CrossRef]
78. Bhattacharyya, R.; Misra, A.; Sandeep, K. Photovoltaic Solar Energy Conversion for Hydrogen Production by Alkaline Water Electrolysis: Conceptual Design and Analysis. *Energy Convers. Manag.* **2017**, *133*, 1–13. [CrossRef]
79. Khalilnejad, A.; Riahy, G. A Hybrid Wind-PV System Performance Investigation for the Purpose of Maximum Hydrogen Production and Storage Using Advanced Alkaline Electrolyzer. *Energy Convers. Manag.* **2014**, *80*, 398–406. [CrossRef]
80. Badwal, S.P.S.; Giddey, S.S.; Munnings, C.; Bhatt, A.I.; Hollenkamp, A.F. Emerging Electrochemical Energy Conversion and Storage Technologies. *Front. Chem.* **2014**, *2*. [CrossRef]
81. Kolli, A.; Gaillard, A.; De Bernardinis, A.; Bethoux, O.; Hissel, D.; Khatir, Z. A Review on DC/DC Converter Architectures for Power Fuel Cell Applications. *Energy Convers. Manag.* **2015**, *105*, 716–730. [CrossRef]
82. Ursúa, A.; Barrios, E.L.; Pascual, J.; San Martín, I.; Sanchis, P. Integration of Commercial Alkaline Water Electrolysers with Renewable Energies: Limitations and Improvements. *Int. J. Hydrog. Energy* **2016**, *41*, 12852–12861. [CrossRef]
83. Zini, G.; Tartarini, P. Wind-Hydrogen Energy Stand-Alone System with Carbon Storage: Modeling and Simulation. *Renew. Energy* **2010**, *35*, 2461–2467. [CrossRef]
84. Akpinar, E.K.; Akpinar, S. An Assessment on Seasonal Analysis of Wind Energy Characteristics and Wind Turbine Characteristics. *Energy Convers. Manag.* **2005**, *46*, 1848–1867. [CrossRef]

85. Babu, N.R.; Arulmozhivarman, P. Wind Energy Conversion Systems-A Technical Review. *J. Eng. Sci. Technol.* **2013**, *8*, 493–507.
86. Douglas, T.G.; Cruden, A.; Infield, D. Development of an Ambient Temperature Alkaline Electrolyser for Dynamic Operation with Renewable Energy Sources. *Int. J. Hydrog. Energy* **2013**, *38*, 723–739. [CrossRef]
87. Kiaee, M.; Cruden, A.; Infield, D.; Chladek, P. Utilisation of Alkaline Electrolysers to Improve Power System Frequency Stability with a High Penetration of Wind Power. *IET Renew. Power Gener.* **2014**, *8*, 529–536. [CrossRef]
88. Varone, A.; Ferrari, M. Power to Liquid and Power to Gas: An Option for the German Energiewende. *Renew. Sustain. Energy Rev.* **2015**, *45*, 207–218. [CrossRef]
89. Parra, D.; Swierczynski, M.; Stroe, D.I.; Norman, S.; Abdon, A.; Worlitschek, J.; O'Doherty, T.; Rodrigues, L.; Gillott, M.; Zhang, X.; et al. An Interdisciplinary Review of Energy Storage for Communities: Challenges and Perspectives. *Renew. Sustain. Energy Rev.* **2017**, *79*, 730–749. [CrossRef]
90. Kiaee, M.; Infield, D.; Cruden, A. Utilisation of Alkaline Electrolysers in Existing Distribution Networks to Increase the Amount of Integrated Wind Capacity. *J. Energy Storage* **2018**, *16*, 8–20. [CrossRef]
91. Hug, W.; Divisek, J.; Mergel, J.; Seeger, W.; Steeb, H. Highly Efficient Advanced Alkaline Electrolyzer for Solar Operation. *Int. J. Hydrog. Energy* **1992**, *17*, 699–705. [CrossRef]
92. Kuroda, Y.; Nishimoto, T.; Mitsushima, S. Self-Repairing Hybrid Nanosheet Anode Catalysts for Alkaline Water Electrolysis Connected with Fluctuating Renewable Energy. *Electrochim. Acta* **2019**, *323*, 134812. [CrossRef]
93. Djafour, A.; Matoug, M.; Bouras, H.; Bouchekima, B.; Aida, M.; Azoui, B. Photovoltaic-Assisted Alkaline Water Electrolysis: Basic Principles. *Int. J. Hydrog. Energy* **2011**, *36*, 4117–4124. [CrossRef]
94. Artuso, P.; Zuccari, F.; Dell'Era, A.; Orecchini, F. PV-Electrolyzer Plant: Models and Optimization Procedure. *J. Sol. Energy Eng.* **2010**, *132*, 031016. [CrossRef]

Article

Pore Network Simulation of Gas-Liquid Distribution in Porous Transport Layers

Nicole Vorhauer [1,*], Haashir Altaf [1], Evangelos Tsotsas [1] and Tanja Vidakovic-Koch [2]

1 Institute of Process Engineering, Otto von Guericke University Magdeburg, 39106 Magdeburg, Germany
2 Max-Planck-Institute for Dynamics of Complex Technical Systems Magdeburg, 39106 Magdeburg, Germany
* Correspondence: nicole.vorhauer@ovgu.de; Tel.: +48-391-67-51684

Received: 13 July 2019; Accepted: 20 August 2019; Published: 23 August 2019

Abstract: Pore network models are powerful tools to simulate invasion and transport processes in porous media. They are widely applied in the field of geology and the drying of porous media, and have recently also received attention in fuel cell applications. Here we want to describe and discuss how pore network models can be used as a prescriptive tool for future water electrolysis technologies. In detail, we suggest in a first approach a pore network model of drainage for the prediction of the oxygen and water invasion process inside the anodic porous transport layer at high current densities. We neglect wetting liquid films and show that, in this situation, numerous isolated liquid clusters develop when oxygen invades the pore network. In the simulation with narrow pore size distribution, the volumetric ratio of the liquid transporting clusters connected between the catalyst layer and the water supply channel is only around 3% of the total liquid volume contained inside the pore network at the moment when the water supply route through the pore network is interrupted; whereas around 40% of the volume is occupied by the continuous gas phase. The majority of liquid clusters are disconnected from the water supply routes through the pore network if liquid films along the walls of the porous transport layer are disregarded. Moreover, these clusters hinder the countercurrent oxygen transport. A higher ratio of liquid transporting clusters was obtained for greater pore size distribution. Based on the results of pore network drainage simulations, we sketch a new route for the extraction of transport parameters from Monte Carlo simulations, incorporating pore scale flow computations and Darcy flow.

Keywords: pore network model; Monte Carlo simulation; drainage invasion; porous transport layer; clustering effect; water electrolysis

1. Introduction

Pore network models (PNMs) are discrete mathematical models that are basically used to simulate and predict pore scale processes. Different types of PNMs are generally distinguished. These are (i) PNMs for quasi-static drainage invasion processes [1–4], (ii) PNMs for quasi-static imbibition invasion processes [1,3–5], (iii) PNMs of drainage with phase transition and diffusion of the vapor (especially applied in drying research) [6–8], and (iv) PNMs for the computation of dynamic pore scale fluid flow [9–13]. While the first three approaches usually assume quasi-static invasion of the pore space, the fourth approach considers viscous flow of the liquid phase and dynamic invasion of the pore space. There are several examples that combine (i) and (ii) [3,4,14]. There also exists a number of PNMs that additionally incorporate pore scale mechanisms, such as liquid film flow [15–17] or crystallization [18–20]. Only a few models are available that take into account coupled heat and mass transfer [21,22] or that at least consider the invasion and transport processes under non-isothermal conditions [23–25].

A wide range of PNM applications related to fuel cells is already available in literature, e.g., [26–33]. Major focus of the PN studies related to fuel cells is essentially on liquid water distribution in dependence of the pore structure [28], liquid clustering [29], the thickness of the gas diffusion layer [30], hydrophobicity [27], local variation of wetting properties, condensation-induced liquid water formation inside the pore structure [31,33], ice formation, model validation by X-ray tomography [32], and neutron tomography [34,35].

So far, PNMs are only rarely applied in water electrolysis research. Experimental studies incorporating microfluidic platforms of pore networks (PNs) are presented by Bazylak et al. [36–39]. Their investigations are based on the correlation of the oxygen flow rate with current density. The microfluidic platforms employed in their studies are based on the geometric information obtained from micro tomography measurements of different titanium porous transport layers (PTLs). They showed that the developing gas bubbles penetrate the porous structure of the model PTLs and form continuous fractal gas branches that cover the PN at the breakthrough point. The network covering gas fingers at the breakthrough point are assumed as stable as long as the volume flow rates correlating with the current density are constant (In [38,39], the air flow rates were 1, 5, and 10 $\mu L\ min^{-1}$ with current densities of 1.4, 7.0, and 14 A cm^{-2}, and the liquid flow rates were 5 and 10 $\mu L\ min^{-1}$. The same liquid flow rates and air flow rate of 1 $\mu L\ min^{-1}$ were applied in the numerical simulations in [37]. In [36], higher liquid flow rates of 505 $\mu L\ min^{-1}$ and gas flow rates of 10 to 300 $\mu L\ min^{-1}$ were applied.) It was illustrated that the shape of the penetrating gas fingers and the final saturation of the microfluidic PN with oxygen are dictated by the pore structure when the invasion occurs in a capillary regime. Higher gas saturation values were obtained for lower porosity of the PN and smaller pore and throat sizes (sintered PTL in [39]). However, an explanation for this outcome is not given in the paper. Additionally, only 2-dimensional (2D) microfluidic experiments are presented; as will be discussed in this paper, the relationship between pore size distribution and saturation is different in 3-dimensional (3D) PNs where essentially the interconnectivity of the pore space is higher. However, it is surmised from PN modeling that the pore size distribution (PSD), especially the standard deviation of pore and throat sizes, were different in the three cases studied in [39] (felt, sintered, and foam PTL). The presence of large and small pores results in the competitive invasion of the PTL. This can especially be illustrated for strongly heterogeneous pore structures, e.g., bimodal PSD, [40]. Principally, the invasion process follows the path of least resistance. In the hydrophilic micromodels used in experiments in [39], this is the path that follows the lowest invasion pressure thresholds of the neighboring pores; it is thus the route of the largest neighbor pores. In more detail, invasion of the PTL at constant current densities is a process of quasi-static drainage of water. This process occurs in distinct steps. This is designated as the mechanism 'one throat at a time' in [36] and can be simulated with PNMs. To invade any liquid filled pore or throat with water, the pressure inside the gas phase has to overcome a critical invasion pressure threshold that depends on the curvature of the gas-liquid interface. If the pore and throat sizes are distinct, the invasion events occur at distinct pressures. According to this, the pressure curve follows a trend of alternating pressure increase (to achieve the critical invasion pressure thresholds), during which the saturation of the microfluidic network remains constant, and the pressure drops during the sudden invasion (saturation decrease) when the critical invasion pressure is achieved [41]. In [36], it is also shown that the invasion of the PN can continue after the breakthrough of one gas branch if the oxygen flow rate, and thus the gas pressures are increased accordingly.

In this paper we focus on water electrolysis cells with in the sandwich coordinate countercurrent flow of water and oxygen (perpendicular to the catalyst layer and the proton exchange membrane) [42]. Additionally, we consider the situation of high oxygen production rates (current density = 0.6–6 A/cm^2 [43]; oxygen flux < 4 $\mu L\ s^{-1}\ mm^{-2}$). Based on this, we assume plug flow of oxygen and water. In the first approach presented in this paper, we use pore network Monte Carlo simulations (PNMCSs) for the prediction of the invasion of an initially fully water saturated PN with oxygen. We furthermore assume that a stationary distribution of oxygen and water inside the PTL is achieved when the oxygen flow paths connect the catalyst layer, where the oxygen is produced, with the water supply channel, while the water connectivity between both sides is maintained, presuming constant oxygen production rates and constant pressure and temperature along the PTL. As a result of the PNMCSs, we show and discuss the invasion patterns in the moment when the water supply route is disconnected, as well as the impact of isolated single liquid clusters on the relative permeability of oxygen and water. Following parameter estimation concepts, e.g., in drainage of soils [1] or drying [44,45], we propose to estimate relative permeability from the stationary final gas-liquid distribution obtained from the PN drainage simulation.

2. Pore Network Model

A schematic sketch of the PNM is given in Figure 1. Here we consider idealized 3D PNs with PSDs in the range of typical PTLs usually applied for water electrolysis (Figure 2). The application of idealized lattice structures is commonly practiced if the physical mechanisms and pore scale effects are analyzed. More advanced studies base the PN simulations on the real structure of the porous medium (e.g., [36,46]). This is also foreseen in our studies in the next step. However, in this paper we present and discuss the basics of the method and the basic outcomes of the drainage simulation and thus disregard the more expensive (concerning time and computational effort) method.

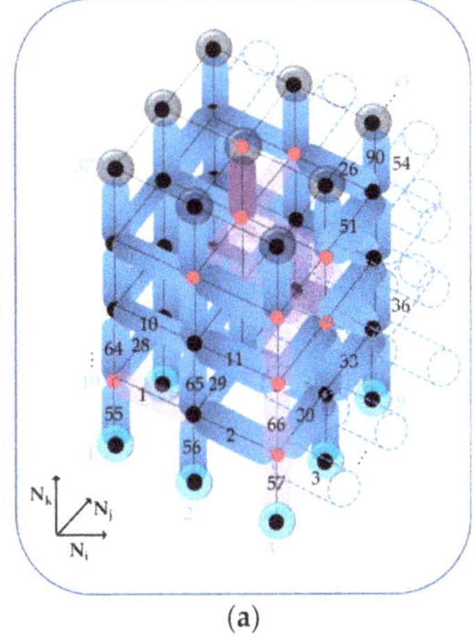

(a)

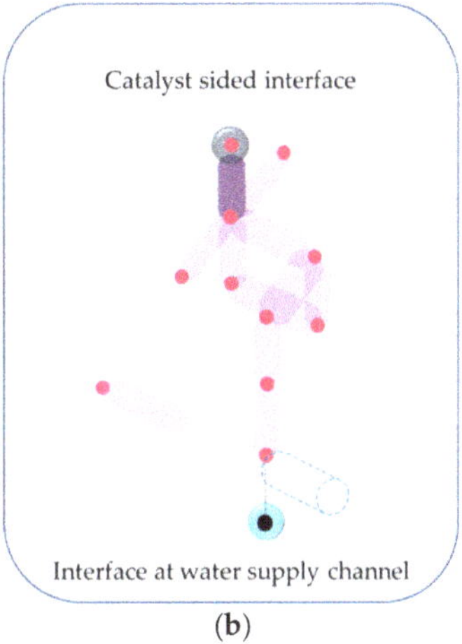

(b)

Figure 1. (**a**) Schematic illustration of the pore network model (PNM) of drainage with pore and throat numbering (pore numbers in blue; throat numbers in black). (**b**) Breakthrough path of oxygen from the catalyst side to the water supply channel. Note the periodic boundary conditions that allow connection of edge throats and pores with each other.

As can be seen in Figure 1a, the PN consists of circular pores, i.e., the sites and cylindrical throats, i.e., the bonds. Both pores and throats can in principal be occupied by either liquid water or gaseous oxygen. The liquid occupied pores are shown in black in Figure 1a; the gas occupied pores are shown in red. Similarly, the liquid occupied throats are shown in blue and the empty throats are shown in red. The pores at the bottom side of the network can be interpreted as water sources as they represent the connection links to the water supply channel. All of these pores are initially saturated with water (black pores in Figure 1). In contrast, pores at the top side of the PN are connected to the oxygen sources inside the catalyst layer at the anode side of the water electrolysis cell and directly supplied with gaseous oxygen (red pores in Figure 1). The water supply route from the water supply channel connected to the bottom of the PN towards the catalyst layer (at the top side) is given by the interconnected blue throats, while the red pores and throats provide distinct routes for the gas phase. Figure 1b shows the path of oxygen through the PN. As can be seen, the oxygen path covers the PN from the catalyst side to the water supply channel in this example. This situation is called a breakthrough. The breakthrough of the gas phase is achieved when one of the bottom pores is occupied with oxygen. On the other hand, the water supply is interrupted once the blue cluster is either completely split up into numerous smaller single clusters or disconnected from either side.

The geometric arrangement of pores and throats and the neighbor relations are specified by the different matrices pnp, tnt, tnp, and pnt. Following the example in Figure 1a, the neighbor relations of the 3D PN as illustrated here are:

$$\mathrm{pnp(pore\,10)} = \begin{bmatrix} \vdots & \vdots & \vdots & \vdots & \vdots & \vdots \\ 1 & 11 & 12 & 13 & 16 & 19 \\ \vdots & \vdots & \vdots & \vdots & \vdots & \vdots \end{bmatrix} \tag{1}$$

$$\mathrm{tnp(throat\,1)} = \begin{bmatrix} 10 & 11 \\ \vdots & \vdots \end{bmatrix} \tag{2}$$

$$\mathrm{pnt(pore\,10)} = \begin{bmatrix} \vdots & \vdots & \vdots & \vdots & \vdots & \vdots \\ 1 & 3 & 28 & 34 & 55 & 64 \\ \vdots & \vdots & \vdots & \vdots & \vdots & \vdots \end{bmatrix} \tag{3}$$

$$\mathrm{tnt(throat\,1)} = \begin{bmatrix} 2 & 3 & 28 & 29 & 34 & 35 & 55 & 56 & 64 & 65 \\ \vdots & \vdots & \vdots & \vdots & \vdots & \vdots & \vdots & \vdots & \vdots & \vdots \end{bmatrix} \tag{4}$$

From this, it follows that the coordination number of the PN is 6. Note that we apply periodic boundary conditions. This means that the pores at all lateral sides are connected to each other in order to reduce confinement effects. While the radius distribution of the throats and pores stochastically vary in the PNMCSs, the other geometrical parameter, such as throat length and the neighbor relations, are kept constant (Table 1). The PSDs of pores and throats in the PN simulations presented below are illustrated in Figure 2.

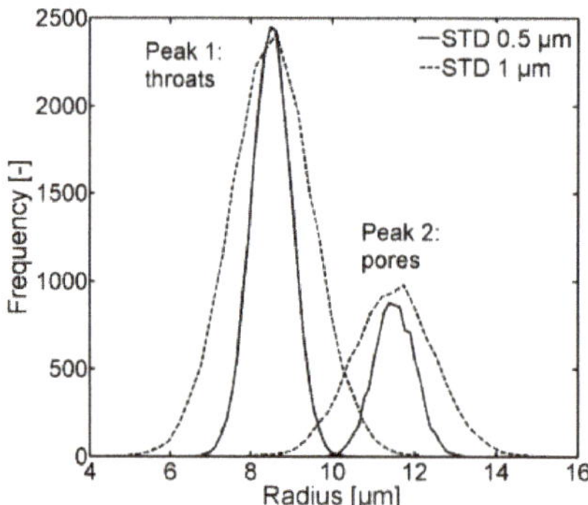

Figure 2. Representative pore size distribution (PSD) of the pore networks (PNs) studied in the pore network Monte Carlo simulations (PNMCSs) in Section 3.1. The PN with low standard deviation (STD) of pore and throat sizes is denoted STD 0.5 μm, and the PN with a higher standard variation is denoted STD 1 μm (refer to discussions below in Section 3.1). The first peaks correspond to the sizes of throats and the second peaks to the pore sizes, respectively. Note that the overlap of pore and throat sizes is greater for a higher standard deviation. The porosity of both PNs is 21%.

Table 1. Simulation parameters.

Parameter	Value
Network size	30 × 30 × 10
Pore number	9000
Throat number	22,500
PTL temperature T	50 °C
Cell pressure P	10 bar
Contact angle θ	0°
Throat length L_t	27 μm
Lattice spacing L	50 μm
Thickness of the PN	450 μm
Porosity	21%

We assume that the oxygen is homogeneously distributed along the surface of the PN and neglect spatial and temporal pressure fluctuations. Instead, we assume a constant oxygen supply rate. Based on this, we anticipate plug flow of oxygen and water and compute the invasion of the PN based on the Young–Laplace equation, following the concepts in [8,47,48]:

$$P_{l,t,p} = P - \frac{2\sigma \cos\theta}{r_{t,p}} \quad (5)$$

The order of invasion thus follows the radius distribution of the throats (with index t) and pores (with index p). Viscous flow is disregarded for the drainage invasion computation. The liquid clusters are labeled based on the throat saturation, taking into account the saturation in neighboring pores. This means that any neighbor throats that contain water belong to the same liquid cluster if the pore between them is also saturated with water (Figure 1). The gas phase is not labeled as it is continuous. Note that liquid clusters that are disconnected from the cluster spanning the PTL from the catalyst side to the water channel side (all blue throats in Figure 1a) are not further invaded. These clusters remain in their original size and can be regarded as transport barriers for oxygen flow (Figure 3). In regard to an efficient PTL mass transfer, it would be affordable to avoid clustering of the liquid phase; relevant issues are also intensively studied e.g., in hydrodynamics of porous geological structures and soils [49]. However, in regard to the optimization of the PTL, several aspects interact with each other, including mass transfer, heat transfer, and electrical conductivity [50].

The PNMCSs presented here are restricted to the computation of the point when the fragmented clusters are not further invaded. In most cases, the fragmented clusters do not span the network as they are either connected to only one of its open sides (top or bottom) or isolated in the center of the PN. The simulation can result in several gas branches penetrating the network from top to bottom. (Note that the gas phase always forms a continuum). This is in good agreement with experimental findings reported in [36].

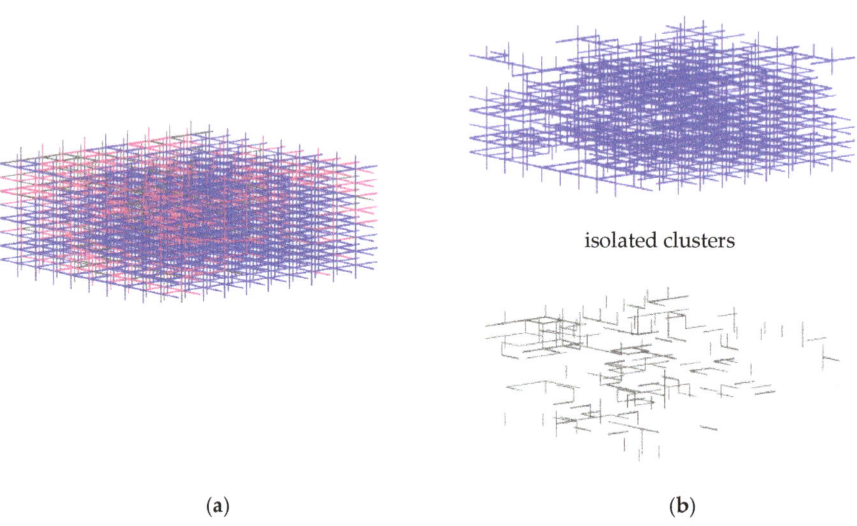

Figure 3. Liquid clustering in a 3-dimensional (3D) PN on the example of a small 10 × 10 × 10 network with homogeneous PSD and the parameters specified in Table 1. The transport barrier clusters (isolated and discrete) are shown in grey (**a**,**b**); the liquid transporting clusters are shown in blue (**a**,**b**); and the continuous gas phase is shown in magenta (**a**). Pores are not shown for reasons of readability.

The computation of the quasi-static invasion profile does not incorporate the solution of mass transfer equations because invasion percolation with trapping is assumed here. Instead, the pore scale fluid transport equations are set-up and solved in the second step based on the stationary invasion patterns from step 1 (Figure 4). For the computation in step 2, the breakthrough invasion patterns, i.e., at disconnection of water transport routes, are used [29,30,37–39]. Only the liquid cluster spanning the PN from top to bottom in the moment before disconnection and the continuous gas phase are considered. The mass transfer through the spanning clusters and the gas phase is computed pore-to-pore, based on the Hagen–Poiseuille equation:

$$\dot{M}_{l,g} = \frac{\rho_{l,g}\pi r_t^{\ 4}}{8\eta_{l,g}L}\left(P_{l,g,1} - P_{l,g,2}\right), \tag{6}$$

for the liquid phase l and the gas phase g. Incompressible, Newtonian viscous flow is assumed (the compression factor of oxygen at operation conditions of P = 10 bar and T = 50 °C is roughly 1). Note that the assumption of the Hagen–Poiseuille flow is a strong model simplification for the pore flow because L ≅ d. A more advanced approach would account for the radial velocity of the flow [51]. In Equation (6), $P_{l,1}$ and $P_{g,1}$ are the liquid and vapor pressures in neighbor pore 1 of a throat (compare with Equation (2)) and $P_{l,2}$ and $P_{g,2}$ are the liquid and vapor pressures in neighbor pore 2 of that throat, respectively. The resulting set of linear equations is then transferred into the matrix notation:

$$P_{l,g} = \mathbf{A}_{l,g}\left(g_{l,g}\right)\backslash b_{l,g} \tag{7}$$

In these equations, $\mathbf{A}_l$ and $\mathbf{A}_g$ represent the matrices of liquid and vapor conductivities of the throats (e.g., refer to [52] for further details), with

$$g_{l,g} = \frac{\rho_{l,g}\pi r_t^{\ 4}}{8\eta_{l,g}L}. \tag{8}$$

Following discussions in [36], conductivities inside pores are not computed as the pores are not interpreted as hydraulic conductors. Validation of this assumption could be further studied by Lattice–Boltzmann simulation, e.g., [53,54].

The boundary conditions for each throat are given in b_l and b_g, which are specified for the pore neighbors 1 and 2 of a throat. Generally:

$$b_{l,g} = g_{l,g} P_{l,g} \tag{9}$$

The boundary conditions for the PN simulation are P_1 at the bottom side of the PN and P_2 at the top side (Figure 4). Solving Equation (7) yields the vapor and liquid pressure fields in the PN. With this, Equation (6) can be solved for each throat. Once the liquid and vapor flow rates through the liquid throats in spanning liquid clusters and the gas throats are known, the overall mass flow rates are computed in step 3 (Figure 4c). From this follows the relative permeability for either the liquid (l) or gas phase (g):

$$K \cdot k_{l,g} = \frac{\eta_{l,g}}{\rho_{l,g}} \frac{\dot{M}_{l,g}}{A} \frac{\Delta z}{\Delta P_{l,g}}. \tag{10}$$

The absolute permeability K is obtained for the same computations but a totally empty (gas permeability) or totally saturated (liquid permeability) network.

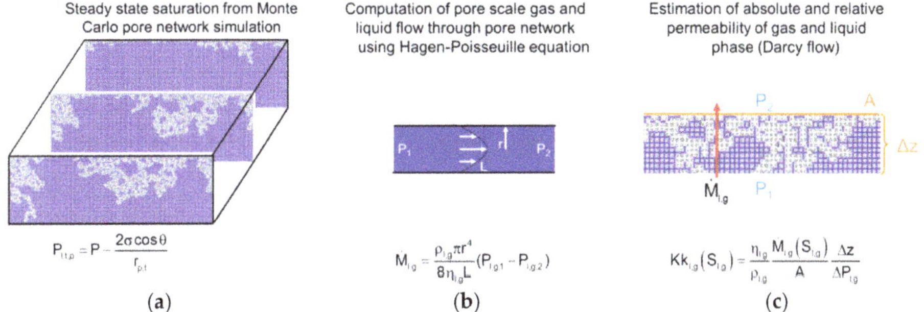

Figure 4. Extraction of efficient transport parameters based on PNMCSs. (**a**) Step 1: Computation of the steady state invasion patterns at the disconnection of the water supply route by PNMCS. (**b**) Step 2: Computation of the pore scale fluid flow based on the patterns obtained from step 1. (**c**) From the flow rates in step 2, the relative and absolute permeabilities of the PN are computed in step 3 on the Darcy scale.

Note that high pressures up to P = 30 bar are postulated as usual operating conditions of water electrolysis cells. It is remarked that the evaporation of water, even at prevailing temperature of 50–80 °C, can be disregarded at such high pressures. Additionally, in the simulations presented here, we generally assumed hydrophilic conditions with cosθ = 1 and constant temperature and pressure as well.

3. PN Simulations and Results

3.1. 3D PNMCS of Drainage

We present here the simulation of one realization of a 3D PN with square lattice and 30 × 30 × 10 pores (Figure 5). The simulation parameters are summarized in Table 1. The overall pressure, as well as the temperature, were kept constant. The lattice spacing between the pores was L = 50 µm and the throat length was kept constant with L_t = 27 µm. The pore and throat sizes varied in the range given in

Figure 2, with standard deviations 0.5 μm and 1 μm. The simulations were repeated 20 times with randomly distributed pore and throat radii within the range given in Figure 2.

According to [55] or [36,46], porosity of the PTL usually varies between 54% and 85%, depending on the kind of the material. Usually, sintered powder has a lower porosity than felt PTLs or foam PTLs. However, the porosity in our simulations was only around 21%. Larger porosities would only be achieved in our PN with greater variation of the pore and throat sizes (i.e., by higher standard deviation) and much shorter throat lengths. To reach the high porosities given in the literature, overlapping of the pore volumes and negligible throat length would have been necessary. This would lead to different invasion effects than currently underlying in the proposed PNM, namely site invasion instead of bond invasion and different flow regimes than those postulated above. (Exemplarily, the porosity could be increased to 43% for throat lengths of 7 μm. $L_t < r_t$ would then require a revision for the assumption of a developed Poiseuille flow inside the throats). Instead, we expect to achieve higher porosities in PNs based on the real porous structure, with various heterogeneities of the pore space (see discussions in Section 4 below). Additionally, due to the relatively long throat length applied in our PN, the thickness is greater than the given values in literature [36,55] where the 10 pore rows correspond to a thickness between 170 μm and 300 μm. Independent of this, the PN simulations presented below nicely illustrate the invasion mechanisms that drive the drainage process. The presented method can be easily adapted to the simulation of the real porous structure.

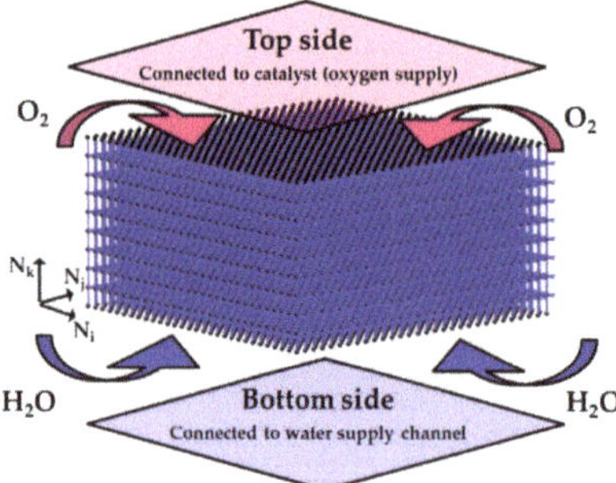

Figure 5. 3D PN under study. Pores are shown in black and throats in blue. The top and bottom pore rows are only vertically connected to their neighbor throats. All pores and throats are initially saturated with liquid.

In these simulations, pores were invaded independently of their throat neighbors. Note that the drainage process is principally a bond invasion process wherefore pores could be invaded together with the throat neighbors [4]. This is also revealed by the experiments presented in [36]. However, as shown in Figure 2 and also represented in Figure 6, the pores are comparably large in our PN with invasion pressure thresholds in the range of the throats. The curves of pore and throat capillary pressures partly overlap and thus indicate the competitive invasion. Apart from that, the invasion pressure thresholds vary linearly with the size of pores and throats if all other parameter are kept constant (Equation (5), Table 1). Figure 6 indicates that the capillary pressures are rather small in the given range of pore sizes, which results in the decrease of liquid pressure (against the gas pressure) of only a few millibars and thus a concave gas-liquid interface.

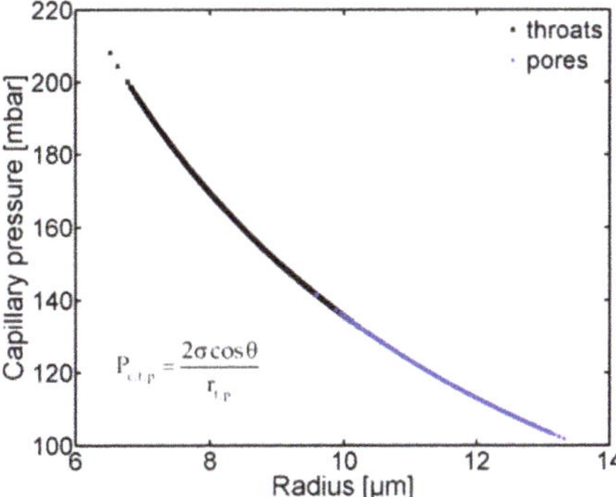

Figure 6. Capillary pressure variation in the studied range of pore sizes (standard deviation of 0.5 μm, Figure 2). Due to the overlapping of the curves the pore invasion is computed independently of the throat invasion.

The simulation results of one PN simulation (with standard deviation 0.5 μm, Figure 2) are summarized below and in Figures 7–9. At first, the capillary pressure curves of the PN simulation are shown in Figure 7. In the hydrophilic drainage process studied here, the available meniscus pore or throat with the lowest capillary pressure is first invaded. Since all surface pores are available with liquid menisci at the start of the simulation, the blue curve initially follows the radius distribution of the surface pores. Once all surface pores are invaded, the capillary curve passes into the trend of the throats, whereas the pore invasion becomes random depending on the distribution of the available meniscus pores. Note that more and more pores become available for invasion when the drainage front widens. However, the blue cloud only represents the pores in liquid clusters spanning the PN, since all other clusters are stationary and not invaded. This holds for the throats analogously. The black curve follows the invasion of the largest throats in the spanning clusters, while the randomly distributed points below the curve correspond again to the PSD of throats in the PN.

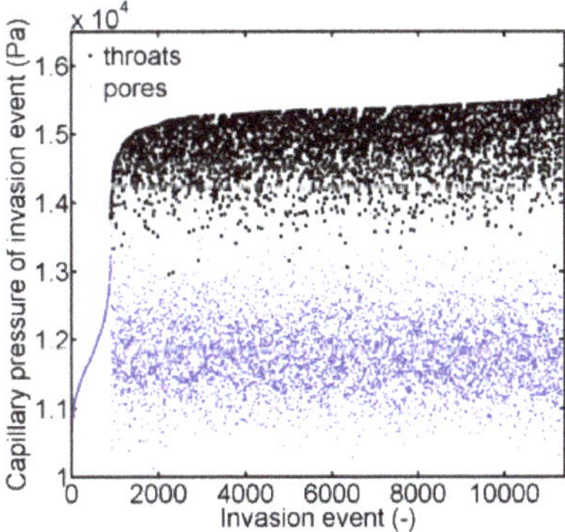

Figure 7. Capillary pressure curves of the drainage invasion process initiated in all surface pores. Note that the small pores and throats with higher capillary pressures (Figure 6) are not invaded.

Figure 8 summarizes the gas-liquid distributions of the PN simulation. At first, the situation a few moments before loss of the liquid connectivity between the vertical throats in the top and bottom row is illustrated in Figure 8a. As can be seen, the liquid phase is multiply disconnected and penetrated by gas branches (in white). The gas phase forms a fractal structure that moves downwards towards the water supply channel. The invasion of the gas phase leads to the formation of isolated liquid clusters that remain behind the front. The size of these clusters is stable as they are not further invaded. It is remarked that the invasion process occurs in all three dimensions in the homogeneous pore structure underlying this simulation (isotropic invasion). This is explained with the equal distribution of liquid pressure (associated with PSDs) throughout the PN, constant wettability, temperature, and pressure conditions. However, it is noted that the horizontal breakthrough occurs much earlier than the vertical breakthrough. This is explained with the oxygen accessibility of the surface pores. In detail, all of the surface pores are initially connected to one liquid-filled throat (downwards) and the gas bulk phase (upwards). Essentially, based on the invasion percolation algorithm and due to the fact that the peak of the PSD of pores is shifted towards larger radii, all pores empty initially before any throat is invaded. This results in the formation of an oxygen front that spans the PN horizontally (the size is roughly NixNj) (Figure 8c,d). According to this, the liquid connectivity between the boundary pores at the top side and the bottom side is already lost at the start of the process. A different situation would be expected if the surface pores were smaller and the invasion pressure were higher.

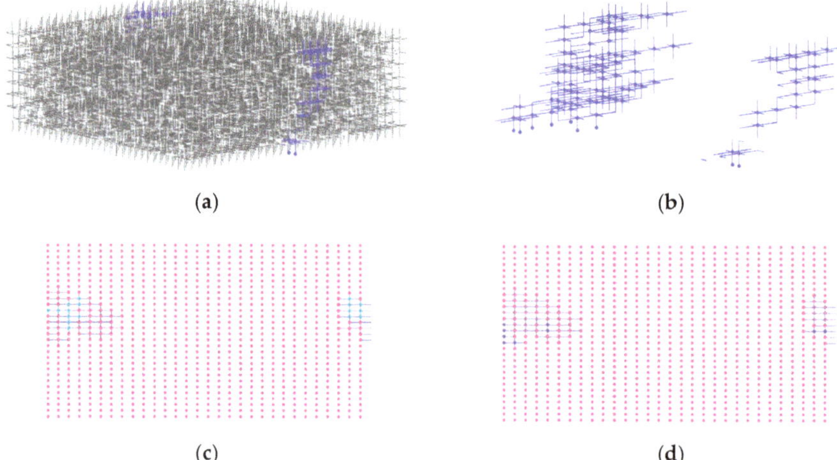

Figure 8. (**a**) PN shortly before loss of connectivity of the liquid supply route through the blue cluster. (**b**) Only the liquid transporting cluster (main cluster) is shown here. Note that due to the applied periodic boundary conditions, the cluster appears at two sides of the PN, while it is disconnected in the center. (**c**) Empty surface pores at the top side (catalyst side). Pores connected to a liquid-filled throat of the spanning cluster are shown in light blue. (**d**) Surface pores at the bottom side (water channel side). Pores connected to the spanning (main) cluster in blue; all other pores in red.

At the moment when the liquid supply route is interrupted (or shortly before that moment), the overall saturation with liquid is S = 0.5996. However, as shown in Figure 8, there exists only one liquid transporting cluster (main cluster). The other liquid-filled throats are shown in grey and are either single throats or part of isolated clusters. The number of clusters at the interruption of liquid connectivity between the surface throats (first and last row of vertical throats) is 933. The maximum cluster size is 624 throats; this is the blue cluster in Figure 8a,b. The mean size of the other clusters is 9.5 throats, thus approximately two orders of magnitude smaller. The liquid transporting cluster contains approximately 5.7417×10^{-3} μL of water. Its volume referred to the total liquid volume contained in the PN at the interruption of the liquid path is only 3%. This reveals the relevance of the pore scale information about the liquid connectivity for the calculation of the liquid permeability. A different situation is expected in presence of wetting liquid films, as will be discussed below. It is also noted that higher permeabilities would be obtained if the drainage simulation would be interrupted already at a higher overall saturation. However, in the simulation presented in Figures 7–9, a breakthrough of the gas phase occurred shortly before disconnection of the main cluster from the top side.

Furthermore, the saturation of the exchange interfaces plays a vital role. The surface and bottom saturations are highlighted in Figure 8c,d. Note that the light blue pores at the top side are already empty. These pores are connected to liquid-filled vertical throats right below the surface. The bottom pores shown in blue are liquid filled. The red pores in these images are either empty (filled with oxygen) or belong to isolated clusters. The ratio of the wetted surface (taking into account the vertical throats inside the first and last layer of the PN) is around 2% at the top surface and around 1% at the bottom side, thus very low. This results in low liquid permeabilities, as will be shown below.

From the distributions shown in Figure 8, the overall liquid saturation S_l of the pore network can be calculated from:

$$S_l = \frac{\sum\limits_{N_t=1}^{22500} V_t \cdot S_t + \sum\limits_{N_p=1}^{9000} V_p \cdot S_p}{\sum V_t + \sum V_p}, \tag{11}$$

with pore and throat saturation S_p and S_t. The overall gas saturation S_g is:

$$S_g = 1 - S_l, \tag{12}$$

accordingly. The liquid and gas saturations are calculated for each network slice (N_k = 1:10) and illustrated as a function of the vertical position ((N_k − 1)·L, see Figure 1) in Figure 9a,b. Figure 9a,b shows the transient saturation profiles of one PN simulation. It reveals that the PN is basically invaded by oxygen from the top side. If one disregards the top and bottom surface layers of the PN (as they have a different overall volume, Figure 1), the saturation of liquid is evenly distributed. This means that isolated liquid clusters homogeneously cover the PN. The liquid confined in the liquid transporting main cluster spanning the PN is illustrated in Figure 9c. The liquid saturation of the main cluster is calculated analog to Equation (11) by only taking into account the liquid inside it. Two different options are compared with the liquid saturation profiles in Figure 9c. These are option 1:

$$S_{MC,V_{tot}} = \frac{\sum\limits_{N_t(N_k)} V_t \cdot S_{t,MC} + \sum\limits_{N_p(N_k)} V_p \cdot S_{p,MC}}{\sum\limits_{N_t(N_k)} V_t + \sum\limits_{N_p(N_k)} V_p}, \tag{13}$$

saturation on the base of the total void volume contained in slice N_k; and option 2:

$$S_{MC,V_l} = \frac{\sum\limits_{N_t(N_k)} V_t \cdot S_{t,MC} + \sum\limits_{N_t(N_k)} V_p \cdot S_{p,MC}}{\sum\limits_{N_t(N_k)} V_t \cdot S_t + \sum\limits_{N_p(N_k)} V_p \cdot S_p}, \tag{14}$$

saturation on the base of the total liquid volume contained in that slice. Note that only the profiles for S = 1 till S = 0.62 are shown here, because the main cluster splits into two clusters at S = 0.62. The agreement of the curves in Figure 9a,c is good at the start of drainage (when the overall saturation S is high), indicating that initially nearly all the liquid is contained in one cluster. At later stages of the invasion process, a gradient develops. Comparison of the blue and black lines in Figure 9c with the red curves reveals that the center of the volume of the main cluster is located at the bottom side of the PN, thus closer to the water supply channel, whereas connectivity to the top side is lost already at the very beginning of the drainage (when all surface pores are invaded). The difference between the black curves and the red curves is related to the liquid volume contained in single clusters. As can be seen, more single clusters are found inside the upper region of the pore network (i.e., at higher values of (N_k − 1)·L). This reveals that the main cluster is traveling through the PN, leaving the isolated clusters behind its front. From this it can be expected that the relative permeability of the liquid phase continuously decreases with progress of invasion, while the oxygen permeability is 0 as long as the breakthrough of the gas phase is not observed (Section 3.2). This outcome might be an indicator for the relevance of the thickness of the PTL for water exchange processes in water electrolysis cells [56]. According to this, thinner PTLs might be more convenient in terms of the water supply routes. However, further simulations are necessary to proof the consistency of this assumption. This finding is essentially an explanation for the low permeability of the liquid phase, as discussed below.

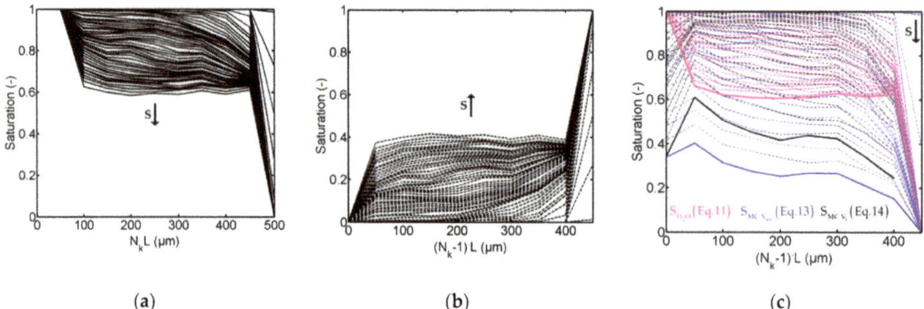

Figure 9. (**a**) Transient liquid saturation profiles of the one PN simulation shown in Figures 7 and 8. (**b**) The according gas saturation profiles. The catalyst side is found on the right (corresponding to the top of PN) and the water supply channel is found on the left (corresponding to the bottom of PN). The saturation profiles are shown for overall liquid saturation S = [1:0.61] (the final saturation is S = 0.6). (**c**) Saturation profiles of the liquid transporting clusters for S = [1:0.62].

It is remarked that different simulation results are obtained for a higher standard deviation (1 μm) (Figure 10). The liquid saturation is still homogeneously distributed throughout the network, but the level of the final slice saturation is lower, namely S = 0.575 (in Figure 9 S = 0.6). Moreover, the main liquid cluster is larger and appears more dense. It contains 8% of the total liquid volume. This reflects the importance of studying the impact of the PSD rather than the impact of porosity, which is around 21% in both PNs [57].

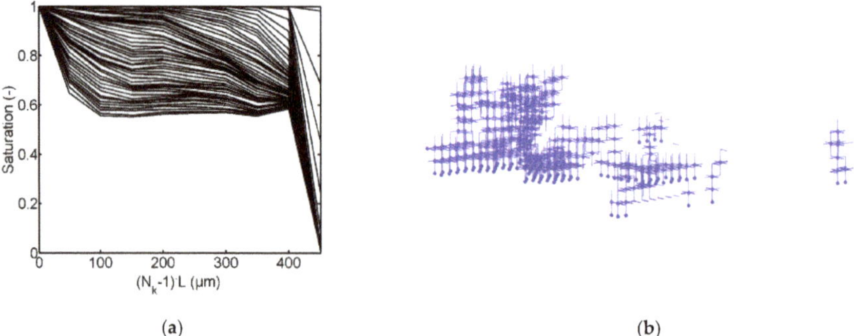

Figure 10. (**a**) Transient saturation profiles of PN with a standard deviation of 1 μm. The saturation profiles are shown for overall liquid saturation S = [1:0.58] (the final saturation is S = 0.575). (**b**) Main cluster.

Consistency of these results is proven by each 20 repetitions of the PN simulation (PNMCS). The results are illustrated in Figure 11, which shows the steady state saturation profiles at the moment of interruption of liquid transport. It is highlighted that the average saturation in the center of the PN is slightly higher in the network with lower standard deviation. This means that more liquid is contained in isolated clusters in this case, as previously observed.

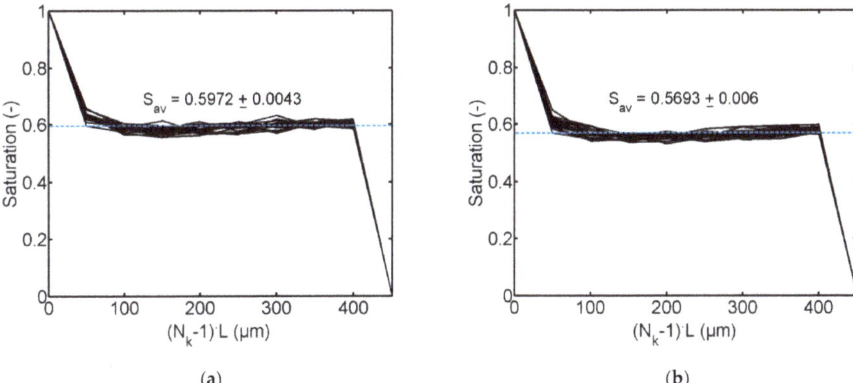

Figure 11. (**a**) Steady state saturation profiles of 20 PN realizations with standard deviation 0.5 μm. (**b**) Steady state saturation profiles of 20 PN realizations with standard deviation 1 μm.

Similar simulation results are obtained if the PN surface is invaded at only one sight. In the simulation shown in Figure 12, the one open pore at the surface is highlighted in light blue. Note that the PN is identical to the one shown in Figures 7–9. The invasion follows a similar route as before and the main cluster spanning the PN at breakthrough is almost identical. The saturation profiles in Figure 12 are obviously different because the surface pore row basically remains saturated. However, the average saturation in the center of the PN is again 0.6 at the end of the process. Isotropy of the invasion process can be shown for such a situation, i.e., when only one pore in the center of the surface area is accessible for oxygen (Figure 13).

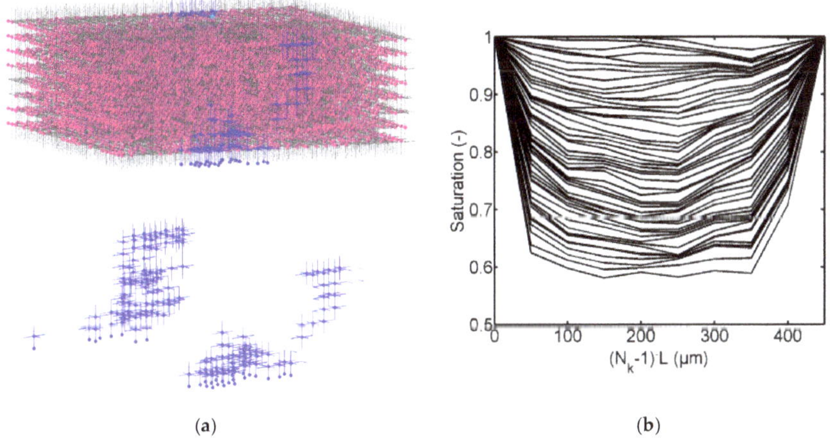

Figure 12. (**a**) PN invasion from one surface pore (shown is the complete PN with gas phase in white, disconnected clusters in grey, and the liquid transporting cluster at the breakthrough of the gas phase in blue). (**b**) Saturation profiles for S = [1:0.64].

Figure 13a plots the label of that in the invasion event invaded vertical (N_k-direction) or horizontal (N_i and N_j-direction) throat (also refer to Figure 1). The three clouds at the start of the process indicate that throats are invaded in each direction (red circles). The breakthrough point is achieved when the throat with the lowest label in each horde is invaded. This is highlighted in Figure 13a. The plot furthermore reveals that, afterwards, horizontal and vertical throats are equally invaded. Figure 13b shows the invasion velocity:

$$\frac{dI'}{dI_{tot}} = \frac{N_{ver,hor}}{N_{tot}}, \quad (15)$$

i.e., the number of invasions in the horizontal and vertical throats (N_{hor} and N_{ver}, respectively) I′ related to the total number of invasions (pores and throats) I_{tot}. According to Figure 13b, more vertical throats are initially invaded, whereas, at later stages of the drainage process, the invasion occurs more often in horizontal throats. This is explained with the PN geometry in Figure 5. It is the main reason for the clustering effect discussed above. In order to prevent the clustering effect, it would be more convenient if the invasion process was anisotropic and occurred mainly in vertical throats. Taking the invasion pressure curve in Figure 6 as a reference, this could be achieved by a gradient in throat sizes (small at the top side, large at the bottom side).

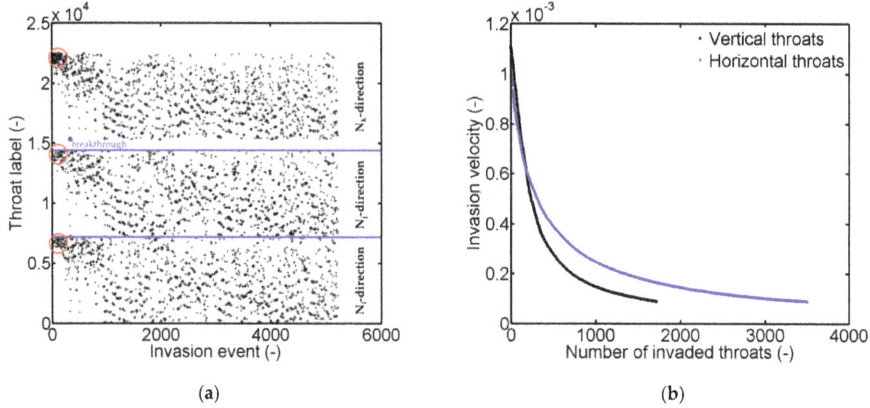

Figure 13. (**a**) Order of invasion on the example of the PN discussed above. According to Figure 1, higher throat labels are associated with the top side of the PN. (**b**) Invasion velocities in vertical and horizontal direction.

It is highlighted that a different invasion behavior would occur if also temperature, pressure, or wettability would spatially vary along the vertical or the horizontal direction. Depending on the situation, anisotropic invasion can also be expected. We plan to illustrate such situations in a forthcoming paper.

3.2. Estimation of Relative Permeabilities

The final gas-liquid phase distribution of the PN drainage simulation is employed for the calculation of gas and liquid permeability. Note that due to the trapping of clusters, the relative permeabilities of the liquid and gas phases cannot be related to each other ($k_l \neq 1 - k_g$).

For calculation of the absolute and relative permeabilities, we have considered the liquid flow through pores in $N_k = 2$ and $N_k = 9$ (Figures 1 and 5) because all surface pores are initially invaded. The pressure gradient is imposed between the pores in these rows. The relative permeability of the liquid phase is related to the total surface area, although only 1–2% of the interfaces are wetted by the liquid spanning cluster. For the PN simulation presented in Figures 7–9, we found an absolute permeability of $K = 1.9639 \times 10^{-12}$ m^2 (liquid phase) and $K = 6.5612 \times 10^{-12}$ m^2 (gas phase) (in [50], similar values are obtained for a porosity of 50%). The relative permeabilities are $k_l = 7.8439 \times 10^{-4}$ for the liquid phase and $k_g = 0.0068$ for the gas phase. The relatively low oxygen permeability is associated to the hindrance by isolated liquid clusters. Higher liquid permeabilities are obtained if the calculations are repeated for higher liquid saturations of the PN, i.e., if the PN is not fully drained (Figure 14). (Note that the gas phase is not penetrating the PN for liquid saturations shown in Figure 14). These values are in good agreement with previous results in [45] and they clearly illustrate the different relative permeabilities of 2D and 3D PNs, which are related to the higher connectivity of the 3D PN and the associated clustering effect. This situation can change if wetting liquid films are considered in the PNM [57]; however, their development and extension depend on the wettability of the surface, temperature, pressure, and process conditions. (In [57] it is mentioned that the wettability alteration of titanium PTLs occurs due to the oxidation of its surface.) Additionally, a heterogeneous pore structure with interpenetrating large and small pores can lead to higher relative permeabilities if the breakthrough occurs in the large pores, whereas the small pores remain saturated and provide the pathway for water transport. The relative permeabilities are thus a function of PSD [57]. The absolute permeability depends on the ratio of pore space to solid, i.e., on the porosity (e.g., [1,50]). However, according to Equation (10), higher absolute permeability does not necessarily increase relative permeabilities.

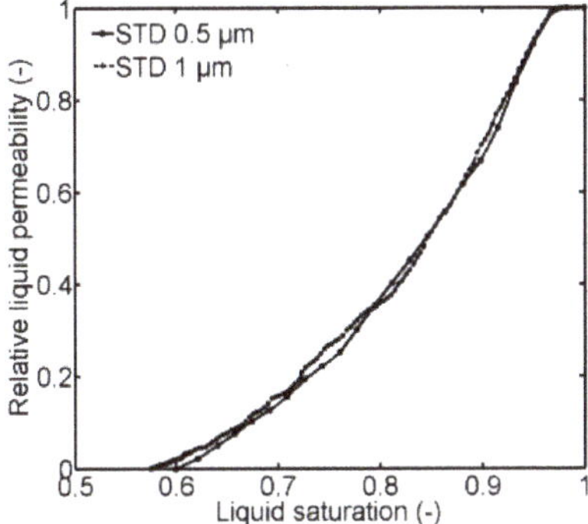

Figure 14. Permeability variation for simulations with low and high standard deviation of PSD.

3.3. Oxygen Production Rate

The decomposition of water occurs if a voltage higher than the thermoneutral voltage is supplied to the electrolysis cell, e.g., U = 1.5–2.2 V [43]. It is anticipated that, at low voltages, oxygen can be solved inside the water that is transported through the PTL towards the catalyst layer. However, if higher oxygen production rates are achieved, the gas can invade the PTL if the pressure in the gas phase is high enough. In [58,59], the different transport regimes for oxygen through the PTL that correlate with the current density of the electrolyzer are shown. In this, dispersed bubbly flow (#1), plug flow (#2), slug flow (#3), churn flow (#4), and annular flow (#5) are distinguished. The latter flow regime is based on wetting liquid films forming along the surface of the pores and sustaining liquid transport between the water flow channel and catalyst layer. For simplicity, we assumed that the current density is always high enough to allow for a plug flow of the gas phase without formation of liquid films; this is option #6 (Figure 15a,b). This assumption is reasonable in the face of the total volume of the pore network of only a few µL (see below). However, if good wettability of the PTL surface is anticipated (hydrophilic properties), liquid films are likely to form along the solid pore surface. These films can enhance liquid transport through invaded pores as they can connect the water supply channel with liquid clusters at the top side of the PTL [15,16,60]. Basically, in presence of liquid films, the previously (and above discussed) isolated clusters would not occur, but rather the complete liquid phase in the PN would be interconnected and participate in liquid transport. This could enhance liquid transport properties significantly. (Exemplarily, in drying processes, liquid films can reduce the overall drying time by orders of magnitude, e.g., [15,17]). The presence of these films strongly depends on the surface roughness, the geometry of pore corners, wettability, temperature, and pressure, as previously mentioned.

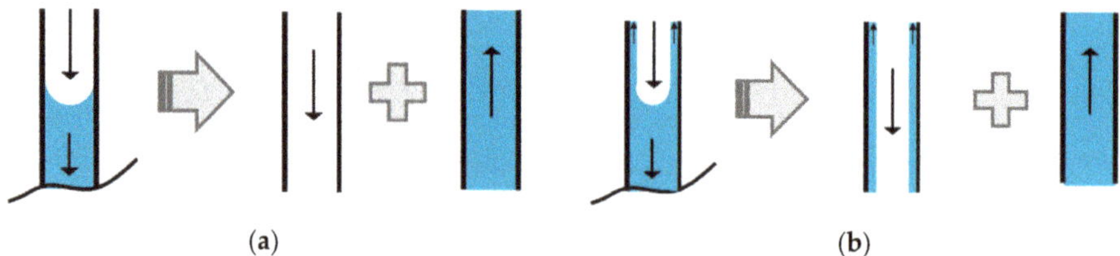

Figure 15. (a) Plug flow invasion of the gas phase (invasion regime #6) completely separating the gas transporting from the liquid transporting pore space. (b) Annular invasion of the gas phase according to [58] (invasion regime #5), leading to annularly wetted pore space transporting gas towards the water supply channel in the center and liquid films transporting water towards the catalyst layer, as well as fully saturated pores and throats.

The drainage algorithm presented above in Section 2 works independently of the current oxygen production rate. We postulate that the oxygen production rate is always high enough to (i) exceed the solubility threshold of oxygen in water (around 40 mg/kg water at 25 °C and 1 bar up to 1 g/kg water at the same temperature and around 25 bar [61]) and to (ii) simultaneously flood the drained channels with oxygen. If this is fulfilled, the invasion process can be seen as fast and is expected to take place in a very short time period.

The oxygen production rate can be calculated using Faraday's law, assuming that the current efficiency of oxygen is equal to 1 and stochiometry

$$2H_2O \rightleftarrows 4H^+ + 4e^- + O_2. \tag{16}$$

Based on Faraday's law:

$$Q = I \cdot t = F \cdot z \cdot N_{O_2}, \tag{17}$$

where z is the number of exchanged electrons (z = 4) and F is Faraday's constant ($F = 9.64853 \times 10^4$ A s mol^{-1}) the moles of electrolyzed water can be correlated with the current of the electrolyzer. From this follows:

$$\frac{dN_{O_2}}{dt} = \dot{N}_{O_2} = \frac{I}{F \cdot z}, \tag{18}$$

the formed molar amount of oxygen per unit time. The volume flux of oxygen can be further calculated from:

$$\dot{v}_{O_2}\left[\frac{m^3}{m^2 s}\right] = \frac{\dot{N}_{O_2}}{A} \cdot \frac{\widetilde{R} \cdot T}{P} = \frac{J}{F \cdot z} \cdot \frac{\widetilde{R} \cdot T}{P}, \tag{19}$$

with current density

$$J = \frac{I}{A} \tag{20}$$

in A m^{-2}. The relationship in Equation (19) is illustrated in Figure 16a; the linear dependence is in agreement with values reported in [39]. Taking the PN from Section 3.1 as a basis, the surface area connected to the catalyst layer is roughly 0.22 mm^2. If it is furthermore assumed that the complete surface area is electrolytically reactive, the oxygen flux in Figure 16a can be converted into a volumetric flow rate:

$$\dot{V}_{O_2}\left[\mu L\ min^{-1}\right] = 0.22\left[mm^2\right]\dot{v}_{O_2}\left[mm\ min^{-1}\right] \cdot 10^3. \tag{21}$$

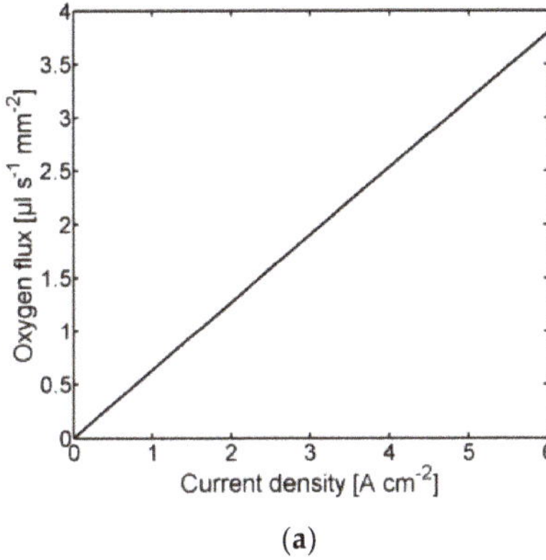

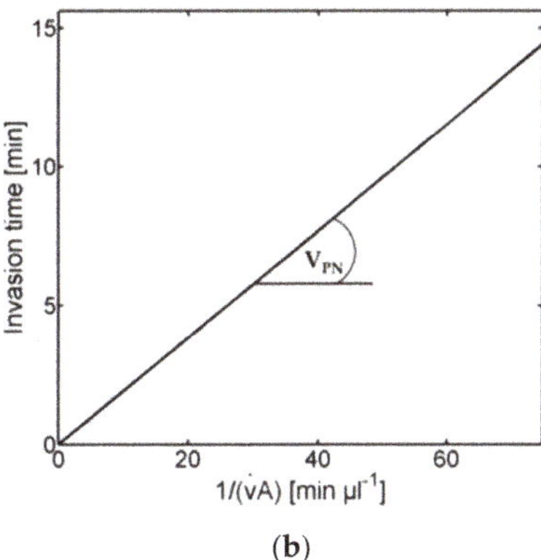

Figure 16. (**a**) Linear dependence of the oxygen volume flux on current density. (**b**) Required invasion time for the constant PN volume V_{PN} (slope) in dependence of the reversed volumetric production flow rate.

Figure 16b illustrates the time dependence of the PN invasion on the reversed volumetric flow $1/\dot{V}_{O_2} = 1/\dot{v}_{O_2}A$ [sμL^{-1}], with the slope being the total volume of the PN in Section 3.1,

$$t = V_{PN}\frac{1}{\dot{v}_{O_2}A}, \tag{22}$$

and A being the total cross section of the PN surface. The total volume of the PN discussed in Section 3.1 is approximately V_{PN} = 0.19 μL; the pore volume is 0.0234 μL and the throat volume is 0.17 μL. In detail, at a current density of 2 A cm^{-2}, the complete PN could be invaded within 0.0675 s, whereas reduction of the current density by factor 1000 increases the invasion time to roughly 1 min. This is in agreement with experiments reported in [36], where, at postulated current densities of 7 A cm^{-2}, the according volume flow rates of the gas phase resulted in invasion times <1 min. Note that always only the dry part of the PN is available for the gas phase. Since this region is much smaller than the total volume, the breakthrough of the gas phase can occur in a shorter time. Additionally, the available cross section for gas invasion is usually lower than the given value in the presence of some liquid pores at the surface (Figure 8).

4. Summary and Discussions

We have presented a method to study the pore scale transport of oxygen and water through the PTL of water electrolysis cells. The method is based on Monte Carlo pore network simulations (PNMCS). It allows simulation of the distribution of the two fluids inside the pore space on a physical base, i.e., without incorporating effective parameters. We have shown that this can be a useful and reliable tool to numerically estimate the pore scale distribution of gas and liquid and the transport coefficients of PTLs, such as relative and absolute permeabilities. More clearly, if these are estimated more accurately based on the proposed method, transport characteristics of PTLs could be predicted more precisely in the future. Moreover, it is highlighted that the discrete method allows correlation of specific transport characteristics with the individual structure of the pore space. The proposed method, therefore, opens up a new route for designing powerful PTLs on customized demand in the future.

The developed PNM basically incorporates invasion percolation rules for hydrophilic drainage with trapping of liquid clusters. Pores and throats are separately invaded in our PN because of their similar invasion pressure thresholds. We assumed plug flow of the liquid and gas phases and negligible film flow. We also assumed that the oxygen production rate is always high enough to homogeneously invade the PN through all surface pores; pressure gradients were disregarded.

The correlation with Faraday's law showed that the small PN can be invaded during less than one minute if the current density is higher than 2 mA cm^{-2}. The invasion process was simulated until the disconnection of the spanning liquid clusters, i.e., when the transport route for water from the top (assumed to be connected to the catalyst layer) to the bottom (assumed to be connected to water supply channel) was interrupted. Due to the discrete character of the model, it was possible to distinguish between isolated clusters, which do not contribute to liquid transport but rather hinder the oxygen flow, and the liquid transporting clusters, which span the network. The possibility to separate single liquid pores and throats depending on their designation as transporting or non-transporting elements particularly reveals the strength of PN modeling and the perspectives for future applications. Different situations were studied. At first, we illustrated relevance of the pore size distribution (PSD). Secondly, we changed the invasion rules of surface pores. For the latter, we allowed invasion of all surface pores in the first situation and restricted invasion to only one pore in the center of the PN in the second situation. We have shown that the final saturation of the PNs with liquid as well as the liquid volume contained in the spanning liquid cluster depends primarily on the PSD; the porosity was kept constant in the situations studied in this paper. Additionally, our simulations revealed that the liquid transporting clusters cover only a very small percentage of the total volume of the PN, whereas most of the liquid phase is disconnected in isolated single clusters if the drainage simulation is continued until disconnection of the water supply route. Additionally, the spanning cluster was travelling through the PN during the drainage invasion and had its center of mass at the bottom side at the moment when liquid connectivity was interrupted. The clustering effect and the travelling spanning cluster are major outcomes of the PN drainage simulation with a significant impact on the liquid and gas permeabilities, which were very low in the studied cases. Furthermore, we found that the spanning liquid clusters providing the transport route for water cover less than 2% of both surfaces (top, bottom). Though this low value can be attributed to the idealized PN structure, the associated low porosity, and the postulated constant boundary conditions, new PN simulations shall further enlighten the major characteristics of mass transfer in PTLs. Therefore, we plan to incorporate micro tomography scans of the PTL in the PNM in the next step. The instrument to be used is a Procon CT alpha with maximum resolution of 0.6 μm, equipped with image processing software, Mavi, from the Fraunhofer Institute for Industrial Mathematics ITWM Kaiserslautern. Following concepts, e.g., described in [1,2,62], we aim to run the PN simulation using the extracted real structure of the porous medium. This will allow more realistic prediction of the gas-liquid distribution for real PTLs as well as study of the role of heterogeneous pore structures, such as bimodal PSDs. Furthermore, microfluidic experiments will be necessary to validate the assumptions of the model. Current open questions concern the limits of the flow regimes (based on discussion in [58,59]), the relevance of liquid flow through corner films [57], the impact of local temperature and wettability variations, as well as the dynamic invasion in the presence of current density fluctuations. Based on [36], we plan to illustrate in more detail the impact of flow regimes (in terms of higher capillary numbers), as well as unsteady operation conditions related to the fluctuation of the current density. In the latter case, we expect multiple redistribution of liquid due to pressure (and eventually also temperature) variations. This implies the application of imbibition invasion rules additionally to drainage rules. In this context, it will also be worth studying if the disconnected liquid clusters can be reconnected and open up more routes for water transport, especially in the situation where liquid films support liquid flow through corners.

Author Contributions: Conceptualization, N.V., E.T., and T.V.-K.; methodology, N.V., E.T., and T.V.-K.; software, N.V. and H.A.; validation, N.V., H.A., and T.V.-K.; formal analysis, N.V. and T.V.-K.; investigation, N.V.; resources, N.V. and E.T.; data curation, N.V. and H.A.; writing—original draft preparation, N.V.; writing—review and editing, T.V.-K.; visualization, N.V.; supervision, N.V.; project administration, N.V., E.T., and T.V-K.; funding acquisition, N.V., E.T., and T.V.-K.

Funding: This research received no external funding.

Conflicts of Interest: The authors declare no conflict of interest.

Abbreviations

Symbol	Parameter (Unit)
A	Area (m^2)
A	Matrix of conductances (-)
b	Vector of boundary conditions (-)
d	Diameter (m)
F	Faraday constant (A s mol^{-1})
g	Conductance (m s)
I	Current (A)
I', I_{tot}	Number of invasion events (-)
J	Area related current (A m^{-2})
k	Relative permeability (-)
K	Absolute permeability (m^2)
L	Length (m)
$\dot{M}$	Flow rate (kg s^{-1})
N	Molar amount (mol), number (-)
N_i, N_j, N_k	Room coordinates (-)
P	Pressure (Pa)
P_c	Capillary pressure (Pa)
pnp	Matrix of pore neighbor relations (-)
pnt	Matrix of pore and throat neighbor relations (-)
Q	Electric charge (C)
r	Radius (m)
$\widetilde{R}$	Universal gas constant (J mol^{-1} K^{-1})
S	Saturation (-)
t	Time (s)
tnp	Matrix of throat and pore neighbor relations (-)
tnt	Matrix of throat neighbor relations (-)
T	Temperature (K)
U	Voltage (V)
V	Volume (m^3)
$\dot{v}$	Velocity (m s^{-1})
$\dot{V}$	Volume flow rate (m^3 s^{-1})
z	Valency number (-), room coordinate (m)
η	Dynamic viscosity (Pa s)
θ	Contact angle (°)
ρ	Density (kg m^{-3})
σ	Surface tension (N m^{-1})
Subscripts	
1,2	Pore 1 or 2
av	Average value
g	Gas phase
hor	Horizontal throats
k	Slice index/number
l	Liquid phase
p	Pore
PN	Pore network
MC	Main cluster
t	Throat
tot	Total
V	Volume related
ver	Vertical throats

Abbreviations

PN	Pore network
PNM	Pore network model
PNMCS	Pore network Monte Carlo simulation
PSD	Pore size distribution
PTL	Porous transport layer

References

1. Blunt, M.J. *Multiphase Flow in Permeable Media*, 1st ed.; Cambridge University Press: Cambridge, UK, 2017.
2. Bultreys, T.; Van Hoorebeke, L.; Cnudde, V. Multi-scale, micro-computed tomography-based pore network models to simulate drainage in heterogeneous rocks. *Adv. Water Resour.* **2015**, *78*, 36–49. [CrossRef]
3. Niasar, V.J.; Hassanizadeh, S.M.; Pyrak-Nolte, L.J.; Berentsen, C. Simulating drainage and imbibition experiments in a high-porosity micromodel using an unstructured pore network model. *Water Resour. Res.* **2009**, *45*, W02430.
4. Patzek, T.W. Verification of a complete network simulator of drainage and imbibition. *SPE J.* **2000**. [CrossRef]
5. Sun, Y.; Kharaghani, A.; Tsotsas, E. Micro-model experiments and pore network simulations of liquid imbibition in porous media. *Chem. Eng. Sci.* **2016**, *150*, 41–53. [CrossRef]
6. Metzger, T. A personal view on pore network models in drying technology. *Dry. Technol.* **2019**, *37*, 497–512. [CrossRef]
7. Vorhauer, N.; Tsotsas, E.; Prat, M. Drying of thin porous disks from pore network simulations. *Dry. Technol.* **2017**, *36*, 651–663. [CrossRef]
8. Prat, M. Percolation model of drying under isothermal conditions in porous media. *Int. J. Multiph. Flow* **1993**, *1*, 691–704. [CrossRef]
9. Börnhorst, M.; Walzel, P.; Rahimi, A.; Kharaghani, A.; Tsotsas, E.; Nestle, N.; Besser, A.; Kleine Jäger, F.; Metzger, T. Influence of pore structure and impregnation: Drying conditions on the solid distribution in porous support materials. *Dry. Technol.* **2016**, *34*, 1964–1978. [CrossRef]
10. Segura, L.A. Modeling at pore-scale isothermal drying of porous materials: Liquid and vapor diffusivity. *Dry. Technol.* **2007**, *25*, 1677–1686. [CrossRef]
11. Metzger, T.; Tsotsas, E. Viscous stabilization of drying front: Three-dimensional pore network simulations. *Chem. Eng. Res. Des.* **2008**, *86*, 739–744. [CrossRef]
12. Yiotis, A.G.; Stubos, A.K.; Boudouvis, A.G.; Tsimpanogiannis, I.N.; Yortsos, Y.C. Pore-network modeling of isothermal drying in porous media. *Transp. Porous Med.* **2005**, *58*, 63–86. [CrossRef]
13. Prat, M.; Bouleux, F. Drying of capillary porous media with a stabilized front in two dimensions. *Phys. Rev. E* **1999**, *60*, 5647–5656. [CrossRef] [PubMed]
14. Fenwick, D.H.; Blunt, M. Three-dimensional modeling of three phase imbibition and drainage. *Adv. Water Res.* **1998**, *21*, 121–143. [CrossRef]
15. Prat, M. On the influence of pore shape, contact angle and film flows on drying of capillary porous media. *Int. J. Heat Mass Tran.* **2002**, *50*, 1455–1468. [CrossRef]
16. Yiotis, A.G.; Boudouvis, A.G.; Stubos, A.K.; Tsimpanogiannis, I.N.; Yortsos, Y.C. Effect of liquid films on the drying of porous media. *AIChE J.* **2004**, *50*, 2721–2737. [CrossRef]
17. Vorhauer, N.; Wang, Y.; Kharaghani, A.; Tsotsas, E.; Prat, M. Drying with formation of capillary rings in a model porous medium. *Transp. Porous Med.* **2015**, *110*, 197–223. [CrossRef]
18. Rahimi, A.; Metzger, T.; Kharaghani, A.; Tsotsas, E. Discrete modelling of ion transport and crystallization in layered porous media during drying. In Proceedings of the 21th International Drying Symposium (IDS2018), Valencia, Spain, 11–14 September 2018.
19. Rahimi, A.; Kharaghani, A.; Metzger, T.; Tsotsas, E. Pore network modelling of a salt solution droplet on a porous substrate: Imbibition, evaporation and crystallization, In Proceedings of the 20th International Drying Symposium (IDS2016), Gifu, Japan, 7–10 August 2016.
20. Veran-Tissoires, S.; Prat, M. Evaporation of a sodium chloride solution from a saturated porous medium with efflorescence formation. *J. Fluid Mech.* **2014**, *749*, 701–749. [CrossRef]

21. Surasani, V.; Metzger, T.; Tsotsas, E. Drying simulations of various 3D pore structures by a nonisothermal pore network model. *Dry. Technol.* **2010**, *28*, 615–623. [CrossRef]
22. Plourde, F.; Prat, M. Pore network simulations of drying of capillary porous media: Influence of thermal gradients. *Int. J. Heat Mass Tran.* **2003**, *46*, 1293–1307. [CrossRef]
23. Huinink, H.P.; Pel, L.; Michels, M.A.J.; Prat, M. Drying processes in the presence of temperature gradients: Pore-scale modelling. *Eur. Phys. J.* **2002**, *9*, 487–498. [CrossRef]
24. Vorhauer, N.; Tsotsas, E.; Prat, M. Temperature gradient induced double-stabilization of the evaporation front within a drying porous medium. *Phys. Rev. Fluids* **2018**, *3*, 114201. [CrossRef]
25. Vorhauer, N.; Tran, Q.T.; Metzger, T.; Tsotsas, E.; Prat, M. Experimental investigation of drying in a model porous medium: Influence of thermal gradients. *Dry. Technol.* **2013**, *31*, 920–929. [CrossRef]
26. Hinebaugh, J.; Bazylak, A. Condensation in PEM fuel cell gas diffusion layers: A pore network modeling approach. *J. Electrochem. Soc.* **2010**, *157*, 1382–1390. [CrossRef]
27. Shahraeeni, M.; Hoorfar, M. Pore-network modeling of liquid water flow in gas diffusion layers of proton exchange membrane fuel cells. *Int. J. Hydrog. Energy* **2014**, *39*, 10697–10709. [CrossRef]
28. Lee, K.J.; Kang, J.H.; Nam, J.H. Liquid water distribution in hydrophobic gas diffusion layers with interconnect rib geometry: An invasion-percolation pore network analysis. *Int. J. Hydrog. Energy* **2014**, *39*, 6646–6656. [CrossRef]
29. Lee, K.J.; Nam, J.H.; Kim, C.J. Steady saturation distribution in hydrophobic gas-diffusion layers of polymer electrolyte membrane fuel cells: A pore-network study. *J. Power Sources* **2010**, *195*, 130–141. [CrossRef]
30. Lee, K.J.; Kang, J.H.; Nam, J.H.; Kim, C.J. Steady liquid water saturation distribution in hydrophobic gas-diffusion layers with engineered pore paths: An invasion-percolation pore-network analysis. *J. Power Sources* **2010**, *195*, 3508–3512. [CrossRef]
31. Straubhaar, B.; Pauchet, J.; Prat, M. Pore network modelling of condensation in gas diffusion layers of proton exchange membrane fuel cells. *Int. J. Heat Mass Tran.* **2016**, *102*, 891–901. [CrossRef]
32. Agaesse, T.; Lamibrac, A.; Büchi, F.N.; Pauchet, J.; Prat, M. Validation of pore network simulations of ex-situ water distributions in a gas diffusion layer of proton exchange membrane fuel cells with X-ray tomographic images. *J. Power Sources* **2010**, *331*, 462–474. [CrossRef]
33. Aghighi, M.; Gostick, J. Pore network modeling of phase change in PEM fuel cell fibrous cathode. *J. Appl. Electrochem.* **2017**, *47*, 1323–1338. [CrossRef]
34. De Beer, F.; van der Merwe, J.H.; Bessarabov, D. PEM water electrolysis: Preliminary investigations using neutron radiography. *Phys. Procedia* **2017**, *88*, 19–26. [CrossRef]
35. Biesdorf, J.; Oberholzer, P.; Bernauer, F.; Kästner, A.; Vontobel, P.; Lehmann, E.H.; Schmidt, T.J.; Boillat, P. Dual spectrum neutron radiography: Identification of phase transitions between frozen and liquid water. *Phys. Rev. Lett.* **2014**, *112*, 248301. [CrossRef] [PubMed]
36. Lee, C.H.; Hinebaugh, J.; Banerjee, R.; Chevalier, S.; Abouatallah, R.; Wang, R.; Bazylak, A. Influence of limiting throat and flow regime on oxygen bubble saturation of polymer electrolyte membrane electrolyzer porous transport layers. *Int. J. Hydrog. Energy* **2017**, *42*, 2724–2735. [CrossRef]
37. Arbabi, F.; Montazeri, H.; Abouatallah, R.; Wang, R.; Bazylak, A. Three-dimensional computational fluid dynamics modelling of oxygen bubble transport in polymer electrolyte membrane electrolyzer porous transport layers. *J. Electrochem. Soc.* **2016**, *163*, 3062–3069. [CrossRef]
38. Arbabi, F.; Kalantarian, A.; Abouatallah, R.; Wang, R.; Wallace, J.; Bazylak, A. Visualizing bubble flows in electrolyzer GDLs using microfluidic platforms. *ECS Trans.* **2013**, *58*, 907–918. [CrossRef]
39. Arbabi, F.; Kalantarian, A.; Abouatallah, R.; Wang, R.; Wallace, J.S.; Bazylak, A. Feasibility study of using microfluidic platforms for visualizing bubble flows in electrolyzer gas diffusion layers. *J. Power Sources* **2014**, *258*, 142–149. [CrossRef]
40. Vorhauer, N.; Metzger, T.; Tsotsas, E.; Prat, M. Experimental investigation of drying by pore networks: Influence of pore size distribution and temperature. In Proceedings of the 4th International Conference on Porous Media and its Applications in Science, Engineering and Industry, Potsdam, Germany, 17–22 June 2012.
41. Morrow, N.R. Physics and thermodynamics of capillary action in porous media. *Ind. Eng. Chem.* **1970**, *62*, 32–56. [CrossRef]

42. Babic, U.; Suermann, M.; Büchi, F.N.; Gubler, L.; Schmidt, T.J. Critical Review—Identifying critical gaps for polymer electrolyte water electrolysis development. *J. Electrochem. Soc.* **2017**, *164*, 387–399. [CrossRef]
43. Bernt, M.; Gasteiger, H.A. Influence of ionomer content in IrO2/TiO2 electrodes on PEM water electrolyzer performance. *J. Electrochem. Soc.* **2016**, *163*, 3179–3189. [CrossRef]
44. Moghaddam, A.A.; Kharaghani, A.; Tsotsas, E.; Prat, M. Kinematics in a slowly drying porous medium: Reconciliation of pore network simulations and continuum modeling. *Phys. Fluids* **2017**, *29*, 022102. [CrossRef]
45. Vorhauer, N.; Metzger, T.; Tsotsas, E. Extraction of effective parameters for continuous drying model from discrete pore network model. In Proceedings of the 17th International Drying Symposium (IDS 2010), Magdeburg, Germany, 3–6 October 2010; pp. 415–422.
46. Schuler, T.; De Bruycker, R.; Schmidt, T.J.; Büchi, F.N. Polymer Electrolyte Water Electrolysis: Correlating porous transport layer structural properties and performance: Part I. Tomographic analysis of morphology and topology. *J. Electrochem. Soc.* **2019**, *166*, 555–565. [CrossRef]
47. Metzger, T.; Irawan, A.; Tsotsas, E. Influence of pore structure on drying kinetics: A pore network study. *AIChE J.* **2007**, *53*, 3029–3041. [CrossRef]
48. Blunt, M.J.; Jackson, M.D.; Piri, M.; Valvatne, P.H. Detailed physics, predictive capabilities and macroscopic consequences for pore-network models of multiphase flow. *Adv. Water Resour.* **2002**, *25*, 1069–1089. [CrossRef]
49. Lenormand, R. Flow through porous media: Limits of fractal patterns. *Proc. R. Soc. Lond. A* **1989**, *423*, 159–168. [CrossRef]
50. Zielke, L.; Fallisch, A.; Paust, N.; Zengerle, R.; Thiele, S. Tomography based screening of flow field/current collector combinations for PEM water electrolysis. *RSC Adv.* **2014**, *4*, 58888. [CrossRef]
51. Tang, T.; Yu, P.; Shan, X.; Chen, H.; Su, J. Investigation of drag properties for flow through and around square arrays of cylinders at low Reynolds numbers. *Chem. Eng. Sci.* **2019**, *199*, 285–301. [CrossRef]
52. Metzger, T.; Tsotsas, E.; Prat, M. Pore-network models: A powerful tool to study drying at the pore level and understand the influence of structure on drying kinetics. In *Drying Technology, Modern Computational Tools at Different Scales*; Tsotsas, E., Mujumdar, A.S., Eds.; Wiley-VCH: Weinheim, Germany, 2011; Volume 1, pp. 57–102.
53. Zachariah, G.T.; Panda, D.; Surasani, V.J. Lattice Boltzmann modeling and simulation of isothermal drying of capillary porous media. In Proceedings of the 21th International Drying Symposium (IDS2018), Valencia, Spain, 11–14 September 2018.
54. Yiotis, A.G.; Psihogios, J.; Kainourgiakis, M.E.; Papaioannou, A.; Stubos, A.K. A lattice Boltzmann study of viscous coupling effects in immiscible two-phase flow in porous media. *Colloids Surf. A Physicochem. Eng. Asp.* **2007**, *300*, 35–49. [CrossRef]
55. Dhanushkodi, S.R.; Capitanio, F.; Biggs, T.; Merida, W. Understanding flexural, mechanical and physicochemical properties of gas diffusion layers for polymer membrane fuel cell and electrolyzer systems. *Int. J. Hydrog. Energy* **2015**, *40*, 16846–16859. [CrossRef]
56. Siracusano, S.; Di Blasi, A.; Baglio, V.; Brunaccini, G.; Briguglio, N.; Stassi, A. Optimization of components and assembling in a PEM electrolyzer stack. *Int. J. Hydrog. Energy* **2011**, *36*, 3333–3339. [CrossRef]
57. Bromberger, K.; Ghinaiya, J.; Lickert, T.; Fallisch, A.; Smolinka, T. Hydraulic ex situ through-plane characterization of porous transport layers in PEM water electrolysis cells. *Int. J. Hydrog. Energy* **2018**, *43*, 2556–2569. [CrossRef]
58. Ito, H.; Maeda, T.; Nakano, A.; Hwang, C.M.; Ishida, M.; Kato, A.; Yoshida, T. Experimental study on porous current collectors of PEM electrolyzers. *Int. J. Hydrog. Energy* **2012**, *37*, 7418–7428. [CrossRef]
59. Ito, H.; Maeda, T.; Nakano, A.; Hasegawa, Y.; Yokoi, N.; Hwang, C.M.; Ishida, M.; Kato, A.; Yoshida, T. Effect of flow regime of circulating water on a proton exchange membrane electrolyzer. *Int. J. Hydrog. Energy* **2010**, *35*, 9550–9560. [CrossRef]
60. Geistlinger, H.; Ding, Y.; Apelt, B.; Schlüter, S.; Küchler, M.; Reuter, D.; Vorhauer, N.; Vogel, H.J. Evaporation study based on micromodel-experiments: Comparison of theory and experiment. *Water Resour. Res.* **2019**. [CrossRef]

61. Kolev, N.I. *Multiphase Fluid Dynamics 4: Turbulence, Gas Adsorption and Release, Diesel Fuel Properties*, 1st ed.; Springer: Berlin, Germany, 2011.
62. Pashminehazar, R.; Ahmed, S.J.; Kharaghani, A.; Tsotsas, E. Spatial morphology of maltodextrin agglomerates from X-ray microtomographic data: Real structure evaluation vs. spherical primary particle model. *Powder Technol.* **2018**, *331*, 204–217. [CrossRef]

Article

Steady-State Water Drainage by Oxygen in Anodic Porous Transport Layer of Electrolyzers: A 2D Pore Network Study

Haashir Altaf [1], Nicole Vorhauer [1,*], Evangelos Tsotsas [1] and Tanja Vidaković-Koch [2]

1 Institute of Process Engineering, Otto von Guericke University, 39106 Magdeburg, Germany; haashir.altaf@ovgu.de (H.A.); evangelos.tsotsas@ovgu.de (E.T.)

2 Electrochemical Energy Conversion, Max Planck Institute for Dynamics of Complex Technical Systems, 39106 Magdeburg, Germany; vidakovic@mpi-magdeburg.mpg.de

* Correspondence: nicole.vorhauer@ovgu.de; Tel.: +49-391-67-51684

Received: 12 February 2020; Accepted: 19 March 2020; Published: 21 March 2020

Abstract: Recently, pore network modelling has been attracting attention in the investigation of electrolysis. This study focuses on a 2D pore network model with the purpose to study the drainage of water by oxygen in anodic porous transport layers (PTL). The oxygen gas produced at the anode catalyst layer by the oxidation of water flows counter currently to the educt through the PTL. When it invades the water-filled pores of the PTL, the liquid is drained from the porous medium. For the pore network model presented here, we assume that this process occurs in distinct steps and applies classical rules of invasion percolation with quasi-static drainage. As the invasion occurs in the capillary-dominated regime, it is dictated by the pore structure and the pore size distribution. Viscous and liquid film flows are neglected and gravity forces are disregarded. The curvature of the two-phase interface within the pores, which essentially dictates the invasion process, is computed from the Young Laplace equation. We show and discuss results from Monte Carlo pore network simulations and compare them qualitatively to microfluidic experiments from literature. The invasion patterns of different types of PTLs, i.e., felt, foam, sintered, are compared with pore network simulations. In addition to this, we study the impact of pore size distribution on the phase patterns of oxygen and water inside the pore network. Based on these results, it can be recommended that pore network modeling is a valuable tool to study the correlation between kinetic losses of water electrolysis processes and current density.

Keywords: pore network model; drainage invasion; pore size distribution; porous transport layer; electrolysis

1. Introduction

With innovations in the energy sector and a need for clean fuel, research is in progress to exploit the potential of hydrogen as an efficient energy source. Exemplarily, fuel cells can utilize hydrogen to produce electricity, and it can also serve as a fuel for internal combustion engines [1]. For the production of hydrogen, electrolyzer technology serves as a very promising and viable option. The purity of the produced hydrogen can be almost 100 vol % [2]. This way, it can be integrated with other renewable resources to offer a broad range of ecologically clean methods for hydrogen production [3–5]. Shortly, electrolyzers and fuel cells will be able to help alleviate the effects of fossil and nuclear fuel consumption [2,6,7]. This, however, implies efficient performance of electrolyzers.

A trade-off between the production rates and the efficiency of an electrolyzer is still to be met. For this technology to be commercial, the cost and hence the efficiency needs to be optimized. Among the electrolyzer technologies, polymer electrolyte membrane (PEM) electrolyzers have an edge over the other varieties because of high energy efficiency and compact design [8–10]. Power needed for the electrolyzer operation can be obtained from renewable sources too and such systems have also been gaining a lot of attention recently [11]. Polymer electrolyte membrane electrolytic cells (PEMECs) are very common when coupling the technology with other renewable sources like solar or wind energy.

High current densities are necessary in order to obtain high production rates. The problem is that the high current density operation results in a decrease of electrical efficiency. This is for example, reported in [12–16] and it is mainly explained with the kinetic losses associated with the mass transfer resistances through the porous transport layers (PTL) [17]. The counter-flow of the two phases causes hindrance against each other, and this mass transfer resistance causes a decrease in the overall electrolyzer performance [18,19].

It is strongly assumed that the high oxygen production rates achieved at the high current densities allow for the formation of gas bubbles that can invade the PTL and partially drain the water [20,21]. The size of gas bubbles increases with the increase in current density [12,15,22]. This leads to the formation of gas fingers penetrating the PTL from the anode catalyst layer, where the oxygen is produced, to the water supply channel from where the oxygen is removed [8]. The development of these gas fingers obviously results in the reduction of the overall water loading of the PTL. This can severely affect the water permeability, especially if the surface saturation of the PTL with water is significantly reduced by gas-filled pores [23]. On the other hand, the efficiency of the oxygen transport in the opposite direction depends mainly on the tortuosity of the evolving gas branches.

According to the work done by Suerman et al. [24], 25% of the total cell overpotential could be contributed to the mass transport losses and these losses are mostly credited to the oxygen withdrawal from the catalyst surface and the PTLs. Yigit et al. [25] reported that at current densities less than 0.7 A/cm^2, the hydrogen production rates were very low and at values above 1 A/cm^2 the electrical efficiency decreased. Larger pores or high porosity values could easily mitigate this mass transport problem on the side of the gas phase, but it would also decrease the electron transport and affect the efficiency [8]. In contrast, small porosity values would hinder the gas removal and increase the entrapped water amount within the catalyst layer, and thus, decrease the rate of reaction [8]. For this purpose, current research aims at the structural optimization of PTLs with respect to efficient mass and electron transfer.

Various experimental methods [12,14–16,20,26–28] and modeling approaches [29–33] are already established to analyze the key factors of the PTL, such as flow regimes, structure, porosity, pore size distribution (PSD), permeability, corrosion resistance, and electrical conductivity. From these studies, the mass transfer limitations discussed before are generally either assigned to the flow regimes of the gas phase (e.g., in Dedigama et al. [12] and Zhang et al. [34]) or to the structure of the PTL. The latter has been studied in [14,15,26,29] (Table 1). In Ito et al. [15] an experimental study on a 27 cm^2 PEM electrolyzer cell was presented that investigates the influence of porosity and pore diameter. In this study, Ti-felt and Ti-powder prepared PTLs with different porosities and different pore structures were used. According to this study, the optimum pore diameter would be 10 μm, whereas no significant effect was seen for a porosity value greater than 50%. Grigoriev et al. [14] estimated an optimum value of 12–13 μm and 30%–50% as the optimum value of porosity using polarization curves for Ti-sintered PTLs with different properties. Findings presented in Kang et al. [26] in an experimental study based on thin/well tunable PTLs in a conventional cell suggested high porosity values and small pore sizes. Ojung et al. [29] used a semi-empirical model in their investigations to study PEM cell without flow channels. They varied pore sizes between 5–30 μm. They observed a decrease in performance at 5 μm and greater than 11 μm there was no significant improvement shown by their model. They also concluded that in a system without flow channels, porosity would not influence the performance at fixed pore diameter and an optimum value of 60% was observed. Hwang et al. [35] also showed with the experiments on reversible fuel cells using Ti-felt that for mean pore sizes around 30–50 μm porosity is an insignificant factor.

Table 1. Study of the interrelation of mass transfer limitations and porous transport layers (PTL) structure.

Reference No.	Type of Material	Technique Used to Study	Estimated Optimum Pore Size	Estimated Optimum Porosity	Operated Current Density
[14]	Ti-sintered	Experimental	12–13 μm	30–50%	0–1.0 A/cm^2
[15]	Ti (felt + sintered)	Experimental	10 μm	<50%	0–2.0 A/cm^2
[26]	Thin/well-tunable Ti	Experimental	400 μm	70%	0–2.0 A/cm^2
[29]	Ti	Semi-empirical model	5–11 μm	60%	0–5.0 A/cm^2

Besides this, explanations for the structure dependence of the electro activity can also be derived from the consideration of flow regimes. Dedigama et al. [12] studied the flow regime within the PTL using electrochemical impedance spectroscopy (EIS) and thermal imaging. They found a reduction in the mass transfer limitation when passing from the bubble to the slug flow regime. In agreement with that, Zhang et al. [34] observed a decrease in the efficiency with an increase of the mass flow rate of water. Aubras et al. [31] showed that the porosity of the anode PTL affects the non-coalesced bubble regime. According to this study, higher porosity can enhance coalescence of oxygen bubbles and increase the performance of electrolyzer. Han et al. [36] also showed an increase in performance linked to an increase in porosity. In addition to that, more recent studies [33,37] revealed the importance of the interaction of the two involved hierarchical porous structures at the anode electrode, namely the PTL and the catalyst layer. Exemplarily, it is demonstrated by Lee et al. [37] using micromodel experiments that the pore sizes control the burst velocity of gas resulting in the application of a thin, low porosity region at the inlet in order to reduce the gas saturations in the PTL.

The majority of the data suggests that a relatively lower value of pore size (around 10 μm) is favorable and no significant conclusion can be drawn about the porosity value. Some authors suggest a high porosity value but others suggest that higher porosity would result in a slug flow, which can lead to inefficient mass transfer and a decrease in efficiency. On the contrary, coalesced bubble regime is also suggested to enhance the performance. In our view, other properties besides mean pore size and porosity might influence the invasion process more significantly. In this paper, we aim at approaching open questions by means of pore network (PN) modeling. In detail, we will consider the role of PSD, which has a higher significance for the invasion in PTL than porosity.

From the above discussions, it can be concluded that advanced manufacturing processes are required to tune the PTL performance. The reader may refer to [26,38–40] for various examples of PTL production techniques that optimize the structure in terms of electrical efficiency and also in terms of material consumption and costs. According to Lettenmeier et al. [39] vacuum plasma spraying can be used to manufacture a PTL with a gradient in pore size along the thickness. It is possible to obtain an average pore size of 10 μm close to the bipolar plate and 5 μm at the electrode side, with the help of this technique (Figure 1). This coating technique can also be used to alter other properties of the PTLs suiting to the need. In a very recent study, Lee et al. [33] showed the effect of porosity gradient on the performance of the electrolyzer. The low porosity region was towards the membrane side and the high porosity region on the water inlet side (Figure 1). They observed high water permeation despite high oxygen saturations. Mo et al. [38] used electron beam melting (EBM) to mitigate the cost and manufacturing issues of the PTL. They showed 8% improvement in the performance compared to the conventionally woven PTLs. They obtained smooth surfaces at both ends of the PTL, thereby reducing the contact resistance between PTL and catalyst layer. Schuler et. al. [41] verified this impact of the interface between PTL and the catalyst layer clearly in their work.

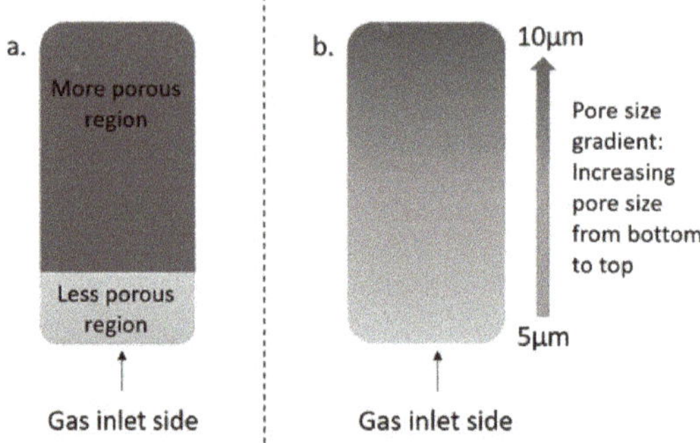

Figure 1. (**a**) Schematic representation of different porosity regions in PTL as studied in [33]. (**b**) Pore size gradient within PTL as in the study from [39].

2. Pore Network Models

The available methods for studying two-phase flow can be divided into continuum and discrete models. The continuum models are usually formulated for macroscale continuous phases employing effective parameters and are thus not suitable for explaining the discrete processes that occur at pore scale. So, in order to gain a deeper understanding of the invasion processes under the action of capillarity, viscous or gravity forces, pore scale models are usually preferred. Pore network models (PNMS) are a type among various pore scale models available, e.g., Lattice Boltzmann models which are mostly available on the scale of one pore. PNMs are discrete models that represent the pore space by a lattice of pores and throats (e.g., [42–46]). In comparison to other methods, which are computationally more expensive in terms of discretization of the physical domain and solving of the governing equations, PNMs can be used to study larger systems computationally more efficiently. Besides fundamental physical studies, PNMs are also used for the parameterization of macroscopic models, e.g., to predict the capillary pressure curve, relative permeability curves, as well as saturation curves [46].

PNMs are generally distinguished between quasi-static and dynamic models [47]. For the simulation of the steady-states in capillary-dominated applications [33,48–51], quasi-static models are used. In the absence of dynamic effects, e.g., driven by viscous forces, the entry capillary pressure of pores and throats controls the displacement of liquid (drainage) or gas (imbibition) in such applications [52]. In this work, the quasi-static model from [46] is applied for the simulation of the drainage of water by oxygen.

The objective of this study is to achieve a fundamental understanding of gas and liquid transport within the PTLs and the pore structure dependence of the invasion patterns. A regular 2D network of pores and throats is used to illustrate the porous PTL. The void space of this network comprises of spherical pores and cylindrical throats. Invasion percolation rules for quasi-static capillary invasion are employed to simulate the displacement of water by oxygen. The displacement mechanism is controlled by the capillary pressure curves of water that are obtained from the Young–Laplace equation:

$$P_l = P_{atm} - \frac{2\sigma cos\theta}{r} \quad (1)$$

where P_l is the liquid pressure, P_{atm} is the atmospheric pressure, σ in kg s^{-2} is the surface tension, θ is the wetting angle, and r is the radius of the channel.

Invasions or displacements occur when they are energetically more favorable. More clearly, the pressure difference between the wetting fluid (water) and the non-wetting fluid (oxygen) leads to the formation of a curvature with a radius depending on the pressure difference. In the case of drainage, the non-wetting phase is the one with higher pressure to be forced through the porous structure.

The incremental increase of the pressure inside the gas phase results in the invasion of the liquid filled pores and throats, with liquid pressures depending on their radius and wettability (Equation (1)). This means, that the stepwise invasion of the interconnected pore space results in the formation of distinct invasion patterns that depend on the PSD and the connectivity of pores and throats.

In such networks, the porosity can be increased basically in two ways, namely by (i) increasing the number of throats of the same dimensions, and (ii) changing the distribution of the throat sizes (Figure 2). In the example illustrated in Figure 3, porosity and mean throat diameter are kept constant and only the distribution varies following the invasion pressure curve computed using Equation (1). As can be seen, the differences in the structural organization of the PN are expected to result in different overall liquid saturations and different gas–liquid distributions [53]. The larger pores and throats are invaded by the gas phase while the smaller ones remain liquid saturated.

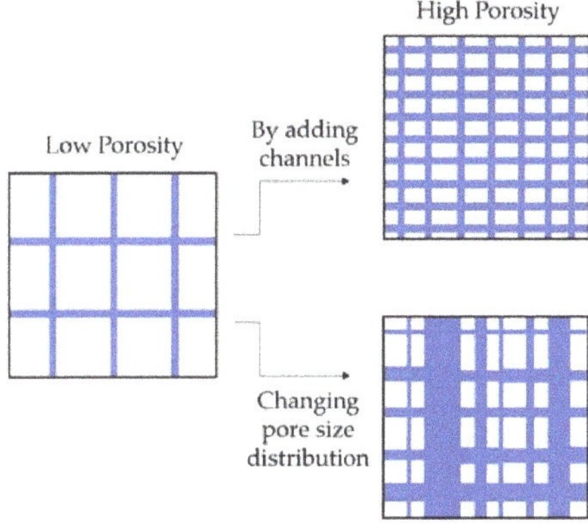

Figure 2. Different throat size distributions with same porosity but different mean throat diameter and standard deviation of the throat size. Solid in white and liquid saturated void space (i.e., pores and throats) in blue.

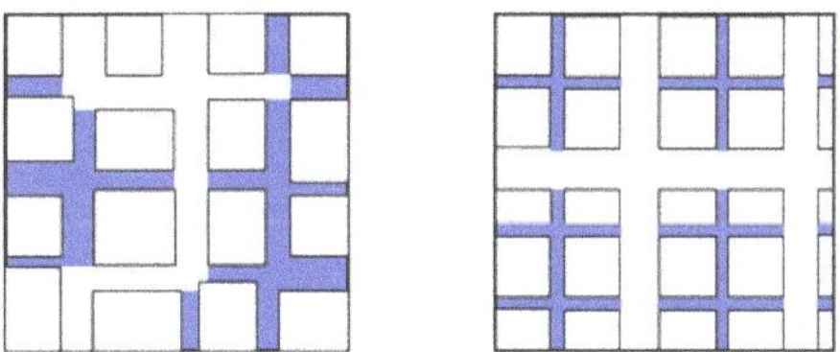

Figure 3. Different saturations for constant porosity and constant mean throat diameter but different organization of the pore network (PN). Solid and empty pores and throats in white and liquid saturated pores and throats in blue.

3. Model Description

The model is comprised of two parts; part one includes the determination of the data structure for the definition of the geometry of the void and solid space, and the second part contains the equations of the drainage algorithm and the cluster labeling (Figure 4). The data structure contains the information about the connections between the pores and throats in the network (Figure 5). This information is used in the drainage algorithm for the stepwise calculation of the successive invasion. The Hoshen–Kopelman algorithm [54,55] is then applied to identify invading and isolated liquid clusters.

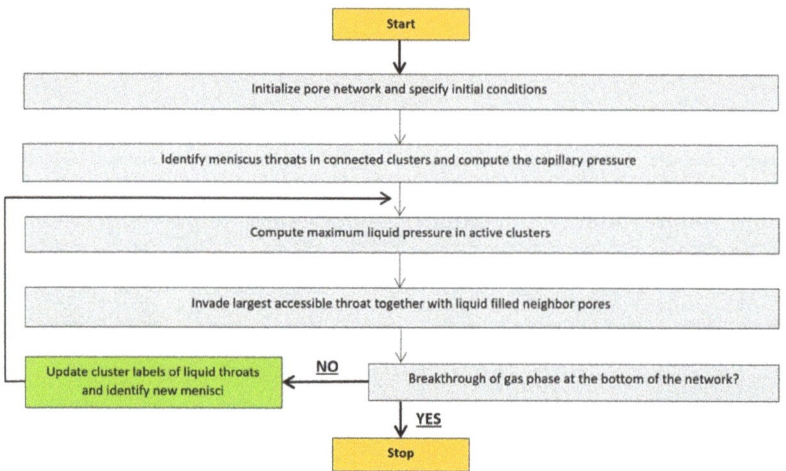

Figure 4. Scheme of the algorithm.

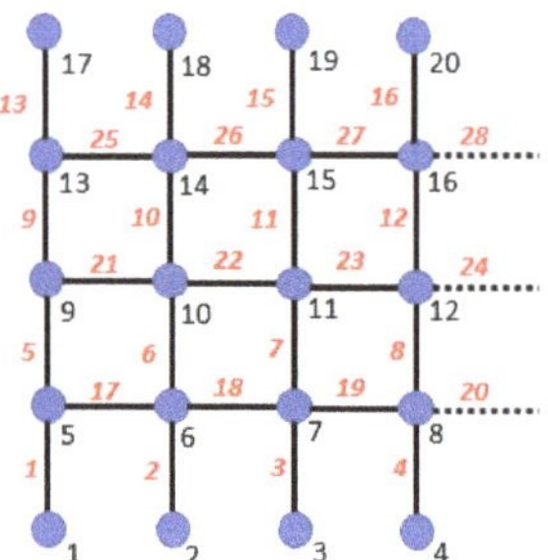

Figure 5. Pore and throat numbering in a 2D PN. The dashed lines illustrate the periodicity at the lateral boundaries.

3.1. Network Generation

Pore and throat radii (r_p, r_t) are randomly distributed around a given average value with defined standard deviations. The geometrical arrangement of throats and neighbor pores with the relevant geometry parameters is illustrated in Figure 6.

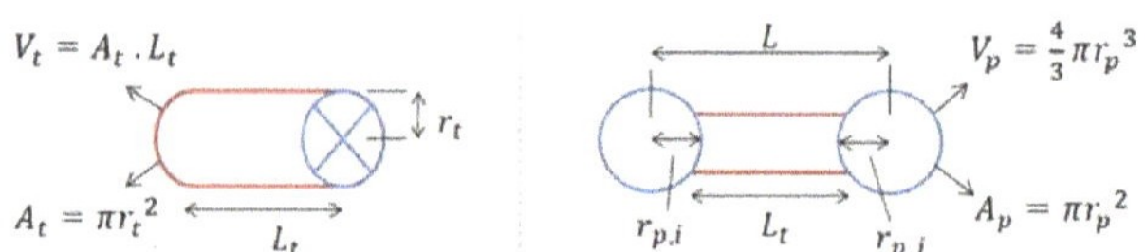

Figure 6. Geometric information about pores and throats.

3.2. Invasion Algorithm

After the geometric parameters are specified, active, i.e., invading clusters with their menisci are identified and the maximum liquid pressure is computed within the active clusters using Equation (1). The algorithm then selects the largest accessible throat or pore for gas invasion following the rules specified in Vorhauer et al. [46]. As invasion proceeds, entrapped clusters are formed, which are permanently isolated due to the incompressibility of the fluids.

3.3. Cluster Labeling

During this process, labeling of liquid clusters is used to identify the pores and throats that are connected to each other and form a pathway for liquid and gas transport. At the start of the invasion process, the network is completely occupied by liquid, and there is only one cluster spanning the whole network and conducting the liquid phase. With initiation of invasion, numerous liquid clusters can form that are distinguished into liquid-conducting clusters (connected to bottom and top side of the PN) and isolated clusters. The Hoshen–Kopelman algorithm [54,55] is used for the labeling of these clusters. Clusters are reviewed and relabeled after each invasion step to update the connections between the pores and throats.

3.4. Model Assumptions

1. Quasi-static drainage invasion in the capillary dominated regime.
2. PN initially saturated with water.
3. No phase transition occurs.
4. Oxygen is injected at the top side and water is removed from the bottom side.
5. There is no mixing or diffusion between the two phases.
6. Viscous, gravity and liquid film flow are neglected.
7. Piston type throat invasion computed based on the Young–Laplace equation.
8. No further invasion occurs after breakthrough of the gas phase.

4. Pore Network Simulation of Microfluidic Experiments

A pore and throat network was conceived using the parameters of microfluidic experiments from Arbabi et al. [20]. The PNM was constructed based on the image data extracted from Figure 7. The experimental image in Figure 7 is then used to compare the flow path of gas within the micromodel qualitatively with our own simulation results. In Figure 7, gas pores and throats are highlighted in blue (pores) and yellow (throats) while liquid pores are in red and liquid throats in black. In this investigation, we are interested if the PNM introduced above is able to predict the experimentally estimated invasion path. It is remarked that invasion is dictated by the interface curvature of pore and throat menisci, wherefore the pore and throat sizes are of interest here. They were determined from the experimental image and transferred into the data structure of the PNM. Although the pore sizes are significantly larger than the throat sizes in this example, pore invasion pressures were not matched with the pore sizes. Instead the pore sizes were randomly adjusted. As can be seen below, the simulation leads already to a very good agreement of the results. However, in a future study, it would be preferable to track experimentally the different invasion pressure thresholds of pores and throats based on the interface curvature of liquid menisci.

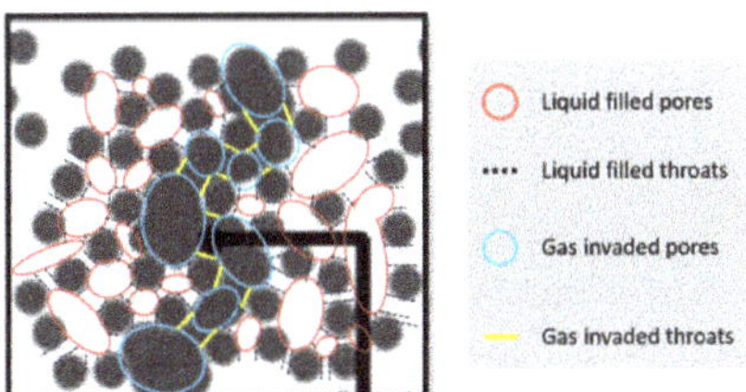

Figure 7. Microfluidic drainage experiment from Arbabi et al. [20] (Reprinted with the permission from Elsevier, 2014). Solids and gas invaded area in black, liquid in white. The PN is identified by the circles and throats. The image shows the steady-state invasion pattern after breakthrough of the gas phase from inlet (**at the bottom**) to water channel (**at the top**).

The data of pore and throat sizes was implemented in the PNM to compute the successive invasion of the PN. The result of the simulation is shown in Figure 8. It is observed that the invasion pattern simulated with the PNM is identical with the experimental image (Figure 7). The perfect agreement reveals the suitability of the model structure and model assumptions and the ability of PNMs to predict the quasi-static drainage patterns. Though, in a future study it would be of interest, if also the stepwise invasion of the PN can be accurately predicted, not only for idealized 2D structures but also in larger and more realistic 3D structures.

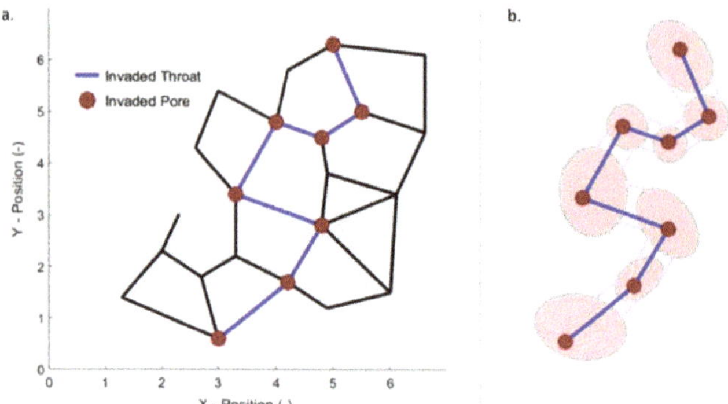

Figure 8. (**a**) PNM simulation result, (**b**) invasion pattern comparison of simulation and experiment result. Liquid-filled throats are shown in black, invaded pores in red, and invaded throats in blue. Liquid-filled pores are not shown.

5. Monte Carlo Simulations

5.1. Impact of Pore Size Distribution in Monomodal PNs

The pore size properties of sintered PTL were used to study the effect of PSD on the invasion patterns and the steady-state saturation of the PTL at breakthrough of the gas phase. For this purpose, a pore network (PN) with the properties summarized in Table 2 were used.

Table 2. Simulation Parameters.

Parameter	Value
Network size (columns and rows)	80 × 30
Temperature	80 °C
Contact Angle	60°
Surface Tension of water	0.0627 N/m
Avg. pore diameter	23 µm
Avg. throat diameter	17 µm
Lattice spacing	50 µm
Avg. throat length	27 µm
Porosity	63%

The standard deviation values were varied from 0.5 µm to 3 µm for a mean throat size of 17 µm (Figure 9), and pore sizes were used with a constant standard deviation value of 2 µm. Monte Carlo simulations were done for each data point so that the given gas saturation values in Figure 10 and Table 3 are an average of 20 simulations. In general, it is observed that the final gas saturation increases at breakthrough with widening of the radius distribution.

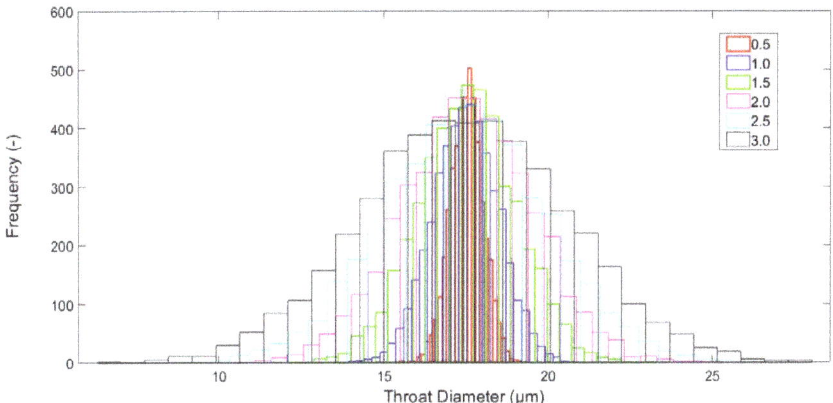

Figure 9. Histograms of pore size distribution (PSD) with varying standard deviation in μm as indicated in the legend.

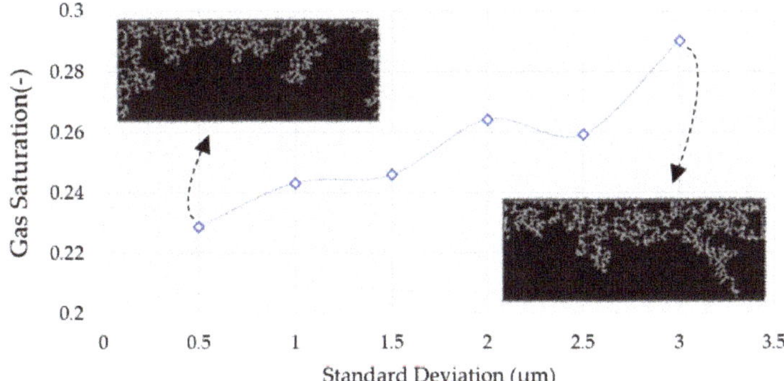

Figure 10. Gas saturation at breakthrough of the gas phase for different PSDs. The mean value of throat sizes is 17 μm.

Table 3. Total void volumes and porosities associated with the gas saturation at breakthrough.

Standard Deviation	0.5 μm	1.0 μm	1.5 μm	2.0 μm	2.5 μm	3.0 μm
Breakthrough gas saturation (%)	22.9	24.3	24.6	26.4	25.6	29
Porosity (%)	63.01	63.05	63.06	63.13	63.14	63.21
Total Void Volume (μL)	0.0178	0.0179	0.0180	0.0185	0.0191	0.0194

The simulation results clearly reveal the impact of PSD on the final distribution of liquid and gas phase. While the average throat size was kept constant, the porosity of the PN increased slightly with increasing standard deviation of throat sizes (Table 3). Thus, following the literature discussions summarized above, it might be anticipated that higher porosities at constant mean throat or pore sizes are a result of an increasing standard deviation of pore and throat sizes in the referenced situations. As shown in Figure 10 and further analyzed below, the variation of PSD affects the invasion and thus the gas saturation. In detail, a higher gas saturation is obtained for broader PSDs. It is to be noted that higher PSDs than presented here can only be studied with a greater mean value of the throat size as will be discussed below.

5.2. Bimodal Pore Size Distributions

The previous results analyze the influence of the PSD on the example of monomodal PNs, i.e., only one peak in the histograms in Figure 9. As can be seen from Table 3, the porosity is only marginally affected in these cases. Due to this, the phase distribution patterns change only slightly in Figure 10. In reality, pore structures often obey bimodal PSDs, i.e., with two peaks in the histogram (Figure 11) and with much stronger impact on porosity. The influence of macro pores is highlighted in Figure 3.

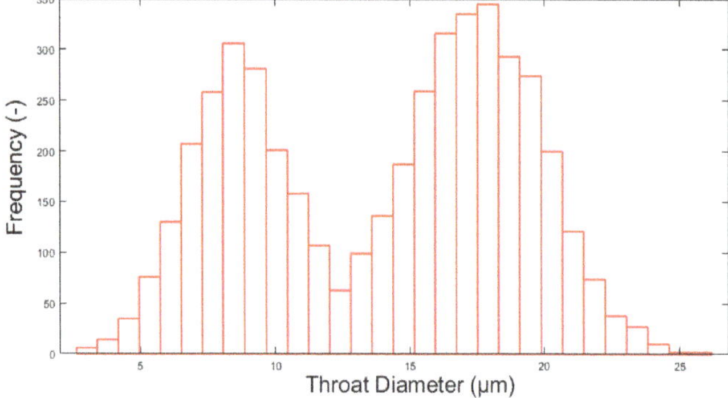

Figure 11. Bimodal throat size distribution with standard deviation 2.0 μm for smaller (the first peak) and 2.5 μm for larger (the second peak) throats.

As can be seen, for the monomodal PN in Figure 12, widespread invasion patterns with a high number of invasions is not achieved. This is in contrast to the bimodal PN in Figure 13. Monte Carlo simulations yielded an average gas saturation of 26% for the monomodal network and 38% for the bimodal network. This shows that widening of the PSD (Figure 9) and the introduction of macro pores (Figure 11), both, result in a change of the invasion process. This change is more significant in the second situation. While capillary fingering with narrow single gas branches is rather favored by narrow PSDs, widening of the invasion front with higher gas saturations is obtained by larger PSDs, and larger porosities. However, it can also be shown that the widening of the front can also be achieved when the porosity is kept constant and only the PSD is adjusted (Figure 13). The monomodal and bimodal networks used in simulations have a constant porosity of 71%. Figures 12 and 13 also show the saturation profiles of these simulations. They reveal the importance of the consideration of the PSD for characterization of the invasion process.

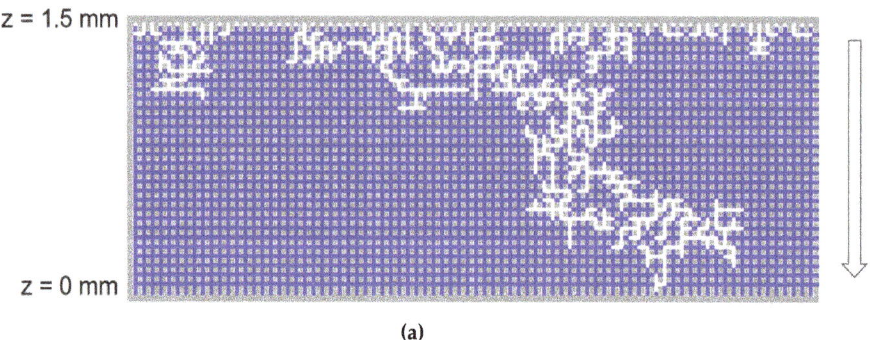

(a)

Figure 12. *Cont.*

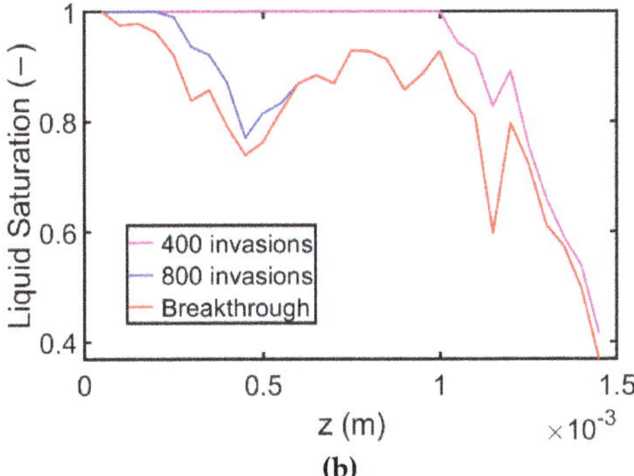

(b)

Figure 12. (**a**) Exemplary invasion patterns of the monomodal PN with porosity 71%. Liquid in blue, gas in white and solid in gray. The arrow indicates the direction of gas invasion. (**b**) Saturation profiles for different overall number of invaded throats and pores achieved during one drainage simulation of the PN with randomly distributed pore and throat sizes.

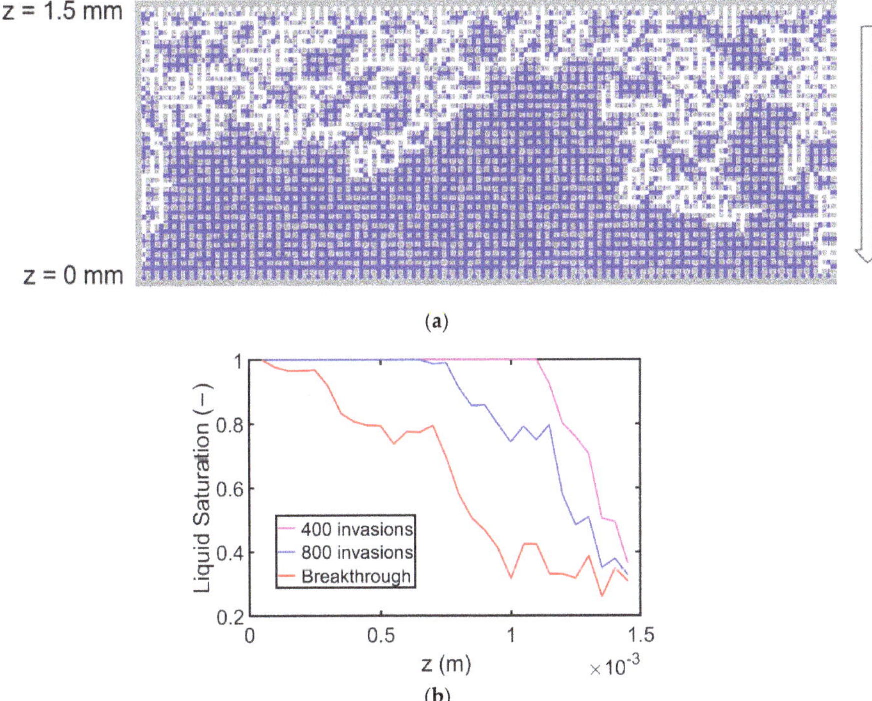

Figure 13. (**a**) Exemplary invasion patterns of the bimodal PN with porosity 71%. Liquid in blue, gas in white and solid in gray. Macro-pores are represented by thicker lines. The arrow indicates the direction of gas invasion. (**b**) Saturation profiles for different overall number of invaded throats and pores from one drainage simulation.

These findings are in very good agreement with literature predictions on drainage invasion regimes [48,56,57]. According to [57], the invasion occurs in the capillary dominated regime, i.e., with rather low injection rate and negligible viscous forces in both fluids. The viscosity ratio of water/air is around 50, and the capillary number is calculated from the ratio of viscous over capillary forces:

$$Ca = \frac{\Delta P_l}{\Delta P_c} \tag{2}$$

with liquid pressure difference ΔP_l and capillary pressure gradient ΔP_c. It depends on the viscosity of the liquid phase η, the wetting curvature by means of surface tension σ, and the velocity of the invading menisci v, as

$$\Delta P_l = \frac{8\eta v L_t}{\bar{r}^2} \tag{3}$$

And

$$\Delta P_c = \frac{2\sigma\sigma_0}{\bar{r}^2} \tag{4}$$

Ref. [58] considering the PSD with mean throat radius $\bar{r}$ and standard deviation σ_0. (In Equation (3), L_t denotes the length of a throat). The conditions for capillary fingering are fulfilled, if the capillary number is below 10^{-4} [48,59]. With the given values for viscosity, surface tension and PSD, this would be achieved if the invasion velocity is below 3.3 mm/min (with liquid viscosity of water $\eta = 10^{-3}$ Pa s and surface tension $\sigma = 0.073$ N/m at 20 °C, throat length $L_t = 50$ µm and standard deviation = 1.5 µm). This is in good agreement with correlations of the invasion time and current density estimated in [46]. Transition to another invasion regime, e.g., stable displacement or viscous fingering [48] would occur if the invasion velocity would be increased or also if the standard deviation of the throat size would be decreased. Such effects are observed in the research of drying of PNs, where sufficiently narrow PSDs can result in stable, i.e., viscosity-dominated invasion [58]. When viscosity comes into play, the width of the invasion front scales with capillary number. However, in the absence of viscosity, the probability of the throat invasion depends on the PSD. In the limit of identical throat sizes, all throats would be equally selected for invasion. In the case of our simulation, where in a such a limit the selection would not be stochastically distributed but ordered by the throat number, the invasion would always occur in the throat with the lowest number (also refer to Figure 4) wherefore consequently this special situation (i.e., no distribution of throat and pore sizes) could not be simulated with the current algorithm. In contrast, when the throat size distribution becomes broader, the invasion follows the path of the least resistance, which results in a more random distribution of the phase patterns, provided that the throat sizes are randomly distributed. Following [60] and [61], the occupation probability decreases with growing Δr.

5.3. Pore Network Simulations of Real Porous Structures

The above discussions are referred to rather artificial porous structures aiming at a more fundamental understanding of the influence of the pore structure on the invasion behavior. In the following, we would like to compare the results for realistic pore structures, extracted from felt, foam and sintered PTLs. The simulation results are compared to literature values from Arbabi et al. [20]. For this purpose, the PN simulations were repeated with a single-entry point (similar as in Figure 7). In more detail, the invasion starts from a single gas pore while all other surface pores are not connected to the oxygen supply. Simulations for each type of PTL were repeated five times with the PSDs shown in Figure 14. From these simulations, representative patterns, i.e., which were most frequently observed, are shown in Figure 15. It is seen that felt and foam PTL show more constricted patterns, while the sintered PTL allows relatively broad patterns with more gas invasions, which is explained well enough by the assumed PSDs for each type (Figure 14).

It is remarked that the PSDs are usually not provided in literature [20]. Due to this, we have selected a standard deviation of throat width so as to catch the gas–liquid phase distributions found for microfluidic experiments in the literature. The agreement of the simulated and the experimental invasion behavior is very well. Also, the results are in line with the discussion of the impact of PSDs above. Namely, the foam exhibits one capillary finger that invades the PN almost straightly. The felt PTL, with a slightly broader PSD (Figure 15), instead reveals a slight widening of the invasion front in horizontal direction. In contrast to that, the sintered PTL allows for a broadening of the invasion front and a significantly higher gas saturation at breakthrough.

In addition to that, it becomes clear from Figure 15 that the effect of liquid clustering occurs in all types of the PTL. The greatest number of isolated liquid clusters is observed in the sintered PTL where the number of invaded pores is higher and the invasion front appears more ramified. In the case of felt and foam PTL, isolated clusters are not seen on the experimental images. This could be because of the small size of the PN and the overall lower number of pore invasions.

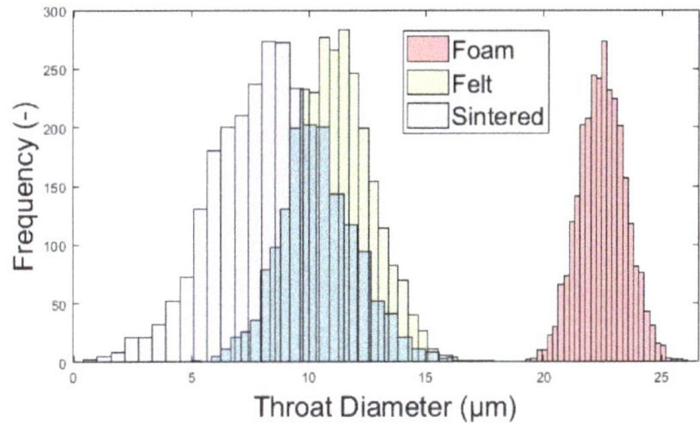

Figure 14. PSDs used for different PTL types with standard deviations: 1.0 μm for foam, 1.7 μm for felt and 2.5 μm for sintered.

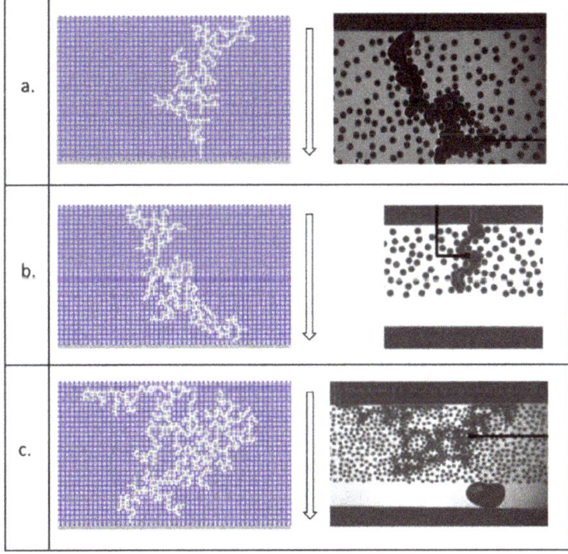

Figure 15. Invasion pattern comparison: (**a**) foam PTL, (**b**) felt PTL, (**c**) sintered PTL (Experimental images from [20] are reprinted with the permission from Elsevier, 2014).

6. Summary and Conclusions

In this study, the applicability of PMNs for the simulation of water drainage from PTL was discussed. The PNM applies invasion percolation rules for hydrophilic drainage. It was shown that the PNM can reliably predict the invasion patterns of 2D microfluidic devices related to water electrolysis studies. In addition to that, the structure dependence of the invasion process was studied using various types of PTLs (foam, felt, sintered). In agreement with experimental data from literature, we could predict different invasion patterns for the three cases. Felt and foam PTLs showed narrow gas fingers while sintered PTL showed widespread invasion front patterns with a higher number of trapped clusters. This observation matches the theoretical investigations based on a Monte Carlo study of PSDs and invasion probability. Based on this, it could furthermore be shown, that the PSD affects the invasion patterns more significantly than the porosity. More clearly, it was revealed that the PSD could be adjusted independently of the porosity and that this resulted in different invasion patterns. The impact of PSD can also be extended towards porous media in fuel cells.

As a next step, the PN will be further enhanced to replicate the local coordination numbers in real PTLs with non-uniform distribution of pores and throats. The purpose would be to simulate mass transfer within the real structure of porous media rather than in an idealized network. These simulations would then be much closer to reality compared to the ones with a fixed coordination number in the network. The effective transport parameters, e.g., permeability, could also then be extracted by solving mass transfer equations pore by pore in a real structure. It would also be important to study the effect of local temperature changes in the system and liquid flow through corner films. Unsteady changes in the current density could also lead to pressure changes. For this reason, the application of imbibition rules along with drainage will become important.

Author Contributions: Conceptualization, N.V. and T.V.-K.; methodology, N.V.; software, H.A.; validation, H.A.; formal analysis, N.V. and T.V.-K.; investigation, H.A.; resources, N.V. and E.T.; data curation, H.A.; writing—original draft preparation, H.A.; writing—review and editing, N.V.; visualization, H.A.; supervision, N.V. and T.V.-K.; project administration, N.V., T.V.-K. and E.T.; funding acquisition, N.V., T.V.-K. and E.T. All authors have read and agreed to the published version of the manuscript.

Funding: This research received no external funding.

Conflicts of Interest: The authors declare no conflict of interest.

Symbols

L	Distance between nodes
L_t	Length of throat
P_c	Capillary pressure
P_l	Liquid pressure
r_p	Pore radius
r_t	Throat radius
$\bar{r}$	Mean radius
v	Velocity of invading menisci
V_p	Volume of pore
V_t	Volume of throat
z	PTL space coordinate
η	Viscosity of liquid phase
θ	Contact angle
σ	Surface tension
σ_0	Standard deviation
Abbreviations	
Ca	Capillary number
PN	Pore network
PNM	Pore network model
PSD	Pore size distribution
PTL	Porous transport layer
Ti	Titanium

References

1. Das, L.M. Hydrogen-fueled internal combustion engines. In *Compendium of Hydrogen Energy*; Elsevier: Amsterdam, The Netherlands, 2016; pp. 177–217.
2. Ursua, A.; Gandia, L.M.; Sanchis, P. Hydrogen Production From Water Electrolysis: Current Status and Future Trends. *Proc. IEEE* **2012**, *100*, 410–426. [CrossRef]
3. Balat, M. Potential importance of hydrogen as a future solution to environmental and transportation problems. *Int. J. Hydrog. Energy* **2008**, *33*, 4013–4029. [CrossRef]
4. Chen, Y.; Hu, X.; Liu, J. Life Cycle Assessment of Fuel Cell Vehicles Considering the Detailed Vehicle Components: Comparison and Scenario Analysis in China Based on Different Hydrogen Production Schemes. *Energies* **2019**, *12*, 3031. [CrossRef]
5. Elgowainy, A.; Gaines, L.; Wang, M. Fuel-cycle analysis of early market applications of fuel cells: Forklift propulsion systems and distributed power generation. *Int. J. Hydrog. Energy* **2009**, *34*, 3557–3570. [CrossRef]
6. Felder, F.A.; Hajos, A. Using Restructured Electricity Markets in the Hydrogen Transition: The PJM Case. *Proc. IEEE* **2006**, *94*, 1864–1879. [CrossRef]
7. Shaw, S.; Peteves, E. Exploiting synergies in European wind and hydrogen sectors: A cost-benefit assessment. *Int. J. Hydrog. Energy* **2008**, *33*, 3249–3263. [CrossRef]
8. Carmo, M.; Fritz, D.L.; Mergel, J.; Stolten, D. A comprehensive review on PEM water electrolysis. *Int. J. Hydrog. Energy* **2013**, *38*, 4901–4934. [CrossRef]
9. Zoulias, E.I.; Glockner, R.; Lymberopoulos, N.; Tsoutsos, T.; Vosseler, I.; Gavalda, O.; Mydske, H.J.; Taylor, P. Integration of hydrogen energy technologies in stand-alone power systems analysis of the current potential for applications. *Renew. Sustain. Energy Rev.* **2006**, *10*, 432–462. [CrossRef]
10. Degiorgis, L.; Santarelli, M.; Calì, M. Hydrogen from renewable energy: A pilot plant for thermal production and mobility. *J. Power Sources* **2007**, *171*, 237–246. [CrossRef]
11. Kosonen, A.; Koponen, J.; Huoman, K.; Ahola, J.; Ruuskanen, V.; Ahonen, T.; Graf, T. Optimization strategies of PEM electrolyser as part of solar PV system. In Proceedings of the IEEE 18th European Conference on Power Electronics and Applications (EPE'16 ECCE Europe), Karlsruhe, Germany, 5–9 September 2016; pp. 1–10.
12. Dedigama, I.; Angeli, P.; Ayers, K.; Robinson, J.B.; Shearing, P.R.; Tsaoulidis, D.; Brett, D.J.L. In situ diagnostic techniques for characterisation of polymer electrolyte membrane water electrolysers—Flow visualisation and electrochemical impedance spectroscopy. *Int. J. Hydrog. Energy* **2014**, *39*, 4468–4482. [CrossRef]
13. Nieminen, J.; Dincer, I.; Naterer, G. Comparative performance analysis of PEM and solid oxide steam electrolysers. *Int. J. Hydrog. Energy* **2010**, *35*, 10842–10850. [CrossRef]
14. Grigoriev, S.A.; Millet, P.; Volobuev, S.A.; Fateev, V.N. Optimization of porous current collectors for PEM water electrolysers. *Int. J. Hydrog. Energy* **2009**, *34*, 4968–4973. [CrossRef]
15. Ito, H.; Maeda, T.; Nakano, A.; Hwang, C.M.; Ishida, M.; Kato, A.; Yoshida, T. Experimental study on porous current collectors of PEM electrolyzers. *Int. J. Hydrog. Energy* **2012**, *37*, 7418–7428. [CrossRef]
16. Mo, J.; Kang, Z.; Yang, G.; Li, Y.; Retterer, S.T.; Cullen, D.A.; Toops, T.J.; Bender, G.; Pivovar, B.S.; Green, J.B., Jr.; et al. In situ investigation on ultrafast oxygen evolution reactions of water splitting in proton exchange membrane electrolyzer cells. *J. Mater. Chem. A* **2017**, *5*, 18469–18475. [CrossRef]
17. Lee, C.H.; Banerjee, R.; Arbabi, F.; Hinebaugh, J.; Bazylak, A. Porous Transport Layer Related Mass Transport Losses in Polymer Electrolyte Membrane Electrolysis: A Review. In Proceedings of the ASME 14th International Conference on Nanochannels, Microchannels, and Minichannels, Washington, DC, USA, 10–14 July 2016. [CrossRef]
18. Ito, H.; Maeda, T.; Nakano, A.; Kato, A.; Yoshida, T. Influence of pore structural properties of current collectors on the performance of proton exchange membrane electrolyzer. *Electrochim. Acta* **2013**, *100*, 242–248. [CrossRef]
19. Lee, C.; Hinebaugh, J.; Banerjee, R.; Chevalier, S.; Abouatallah, R.; Wang, R.; Bazylak, A. Influence of limiting throat and flow regime on oxygen bubble saturation of polymer electrolyte membrane electrolyzer porous transport layers. *Int. J. Hydrog. Energy* **2017**, *42*, 2724–2735. [CrossRef]
20. Arbabi, F.; Kalantarian, A.; Abouatallah, R.; Wang, R.; Wallace, J.S.; Bazylak, A. Feasibility study of using microfluidic platforms for visualizing bubble flows in electrolyzer gas diffusion layers. *J. Power Sour.* **2014**, *258*, 142–149. [CrossRef]

21. Nie, J.; Chen, Y. Numerical modeling of three-dimensional two-phase gas–liquid flow in the flow field plate of a PEM electrolysis cell. *Int. J. Hydrog. Energy* **2010**, *35*, 3183–3197. [CrossRef]
22. Ito, H.; Maeda, T.; Nakano, A.; Hasegawa, Y.; Yokoi, N.; Hwang, C.M.; Ishida, M.; Kato, A.; Yoshida, T. Effect of flow regime of circulating water on a proton exchange membrane electrolyzer. *Int. J. Hydrog. Energy* **2010**, *35*, 9550–9560. [CrossRef]
23. Badwal, S.P.S.; Giddey, S.; Munnings, C. Emerging technologies, markets and commercialization of solid-electrolytic hydrogen production. *WIRE Energy Environ.* **2018**, *7*, e286. [CrossRef]
24. Babic, U.; Suermann, M.; Büchi, F.N.; Gubler, L.; Schmidt, T.J. Critical Review—Identifying Critical Gaps for Polymer Electrolyte Water Electrolysis Development. *J. Electrochem. Soc.* **2017**, *164*, F387–F399. [CrossRef]
25. Yigit, T.; Selamet, O.F. Mathematical modeling and dynamic Simulink simulation of high-pressure PEM electrolyzer system. *Int. J. Hydrog. Energy* **2016**, *41*, 13901–13914. [CrossRef]
26. Kang, Z.; Mo, J.; Yang, G.; Retterer, S.T.; Cullen, D.A.; Toops, T.J.; Green, J.B., Jr.; Mench, M.M.; Zhang, F.-Y. Investigation of thin/well-tunable liquid/gas diffusion layers exhibiting superior multifunctional performance in low-temperature electrolytic water splitting. *Energy Environ. Sci.* **2017**, *10*, 166–175. [CrossRef]
27. Sadeghi Lafmejani, S.; Olesen, A.C.; Kaer, S.K. Analysing Gas-Liquid Flow in PEM Electrolyser Micro-Channels. *ECS Trans.* **2016**, *75*, 1121–1127. [CrossRef]
28. Selamet, O.F.; Pasaogullari, U.; Spernjak, D.; Hussey, D.S.; Jacobson, D.L.; Mat, M. In Situ Two-Phase Flow Investigation of Proton Exchange Membrane (PEM) Electrolyzer by Simultaneous Optical and Neutron Imaging. *ECS Trans.* **2011**, *41*, 349–362.
29. Ojong, E.T.; Kwan, J.T.H.; Nouri-Khorasani, A.; Bonakdarpour, A.; Wilkinson, D.P.; Smolinka, T. Development of an experimentally validated semi-empirical fully-coupled performance model of a PEM electrolysis cell with a 3-D structured porous transport layer. *Int. J. Hydrog. Energy* **2017**, *42*, 25831–25847. [CrossRef]
30. Abdol Rahim, A.H.; Tijani, A.S.; Kamarudin, S.K.; Hanapi, S. An overview of polymer electrolyte membrane electrolyzer for hydrogen production: Modeling and mass transport. *J. Power Sour.* **2016**, *309*, 56–65. [CrossRef]
31. Aubras, F.; Deseure, J.; Kadjo, J.-J.A.; Dedigama, I.; Majasan, J.; Grondin-Perez, B.; Chabriat, J.-P.; Brett, D.J.L. Two-dimensional model of low-pressure PEM electrolyser: Two-phase flow regime, electrochemical modelling and experimental validation. *Int. J. Hydrog. Energy* **2017**, *42*, 26203–26216. [CrossRef]
32. Olesen, A.C.; Rømer, C.; Kær, S.K. A numerical study of the gas-liquid, two-phase flow maldistribution in the anode of a high pressure PEM water electrolysis cell. *Int. J. Hydrog. Energy* **2016**, *41*, 52–68. [CrossRef]
33. Lee, J.K.; Lee, C.; Bazylak, A. Pore network modelling to enhance liquid water transport through porous transport layers for polymer electrolyte membrane electrolyzers. *J. Power Sour.* **2019**, *437*, 226910. [CrossRef]
34. Zhang, H.; Lin, G.; Chen, J. Evaluation and calculation on the efficiency of a water electrolysis system for hydrogen production. *Int. J. Hydrog. Energy* **2010**, *35*, 10851–10858. [CrossRef]
35. Hwang, C.M.; Ishida, M.; Ito, H.; Maeda, T.; Nakano, A.; Hasegawa, Y.; Yokoi, N.; Kato, A.; Yoshida, T. Influence of properties of gas diffusion layers on the performance of polymer electrolyte-based unitized reversible fuel cells. *Int. J. Hydrog. Energy* **2011**, *36*, 1740–1753. [CrossRef]
36. Han, B.; Mo, J.; Kang, Z.; Yang, G.; Barnhill, W.; Zhang, F.-Y. Modeling of two-phase transport in proton exchange membrane electrolyzer cells for hydrogen energy. *Int. J. Hydrog. Energy* **2017**, *42*, 4478–4489. [CrossRef]
37. Lee, C.; Zhao, B.; Abouatallah, R.; Wang, R.; Bazylak, A. Compressible-Gas Invasion into Liquid-Saturated Porous Media: Application to Polymer-Electrolyte-Membrane Electrolyzers. *Phys. Rev. Appl.* **2019**, *11*, 054029. [CrossRef]
38. Mo, J.; Dehoff, R.R.; Peter, W.H.; Toops, T.J.; Green, J.B.; Zhang, F.-Y. Additive manufacturing of liquid/gas diffusion layers for low-cost and high-efficiency hydrogen production. *Int. J. Hydrog. Energy* **2016**, *41*, 3128–3135. [CrossRef]
39. Lettenmeier, P.; Kolb, S.; Sata, N.; Fallisch, A.; Zielke, L.; Thiele, S.; Gago, A.S.; Friedrich, K.A. Comprehensive investigation of novel pore-graded gas diffusion layers for high-performance and cost-effective proton exchange membrane electrolyzers. *Energy Environ. Sci.* **2017**, *10*, 2521–2533. [CrossRef]
40. Kang, Z.; Mo, J.; Yang, G.; Li, Y.; Talley, D.A.; Retterer, S.T.; Cullen, D.A.; Toops, T.J.; Brady, M.P.; Bender, G.; et al. Thin film surface modifications of thin/tunable liquid/gas diffusion layers for high-efficiency proton exchange membrane electrolyzer cells. *Appl. Energy* **2017**, *206*, 983–990. [CrossRef]

41. Schuler, T.; Schmidt, T.J.; Büchi, F.N. Polymer Electrolyte Water Electrolysis: Correlating Performance and Porous Transport Layer Structure: Part II. Electrochemical Performance Analysis. *J. Electrochem. Soc.* **2019**, *166*, F555–F565. [CrossRef]
42. Prat, M.; Agaësse, T. Thin Porous Media. In *Handbook of Porous Media*, 3rd ed.; Vafai, K., Ed.; CRC Press: Boca Raton, FL, USA, 2015; pp. 89–112.
43. Raeini, A.Q.; Yang, J.; Bondino, I.; Bultreys, T.; Blunt, M.J.; Bijeljic, B. Validating the Generalized Pore Network Model Using Micro-CT Images of Two-Phase Flow. *Transp. Porous. Med.* **2019**, *130*, 405–424. [CrossRef]
44. Tranter, T.G.; Gostick, J.T.; Burns, A.D.; Gale, W.F. Pore Network Modeling of Compressed Fuel Cell Components with OpenPNM. *Fuel Cells* **2016**, *16*, 504–515. [CrossRef]
45. Metzger, T. A personal view on pore network models in drying technology. *Dry. Technol.* **2019**, *37*, 497–512. [CrossRef]
46. Vorhauer, N.; Altaf, H.; Tsotsas, E.; Vidakovic-Koch, T. Pore Network Simulation of Gas-Liquid Distribution in Porous Transport Layers. *Processes* **2019**, *7*, 558. [CrossRef]
47. Joekar-Niasar, V.; Hassanizadeh, S.M. Analysis of Fundamentals of Two-Phase Flow in Porous Media Using Dynamic Pore-Network Models: A Review. *Crit. Rev. Environ. Sci. Technol.* **2012**, *42*, 1895–1976. [CrossRef]
48. Sinha, P.K.; Wang, C.-Y. Pore-network modeling of liquid water transport in gas diffusion layer of a polymer electrolyte fuel cell. *Electrochim. Acta* **2007**, *52*, 7936–7945. [CrossRef]
49. Sinha, P.K.; Wang, C.-Y. Liquid water transport in a mixed-wet gas diffusion layer of a polymer electrolyte fuel cell. *Chem. Eng. Sci.* **2008**, *63*, 1081–1091. [CrossRef]
50. Lee, K.-J.; Nam, J.H.; Kim, C.-J. Pore-network analysis of two-phase water transport in gas diffusion layers of polymer electrolyte membrane fuel cells. *Electrochim. Acta* **2009**, *54*, 1166–1176. [CrossRef]
51. Bultreys, T.; Singh, K.; Raeini, A.Q.; Ruspini, L.C.; Øren, P.-E.; Berg, S.; Rücker, M.; Bijeljic, B.; Blunt, M.J. Verifying pore network models of imbibition in rocks using time-resolved synchrotron imaging. *EarthArXiv* **2019**. [CrossRef]
52. Vorhauer, N.; Tran, Q.T.; Metzger, T.; Tsotsas, E.; Prat, M. Experimental Investigation of Drying in a Model Porous Medium: Influence of Thermal Gradients. *Dry. Technol.* **2013**, *31*, 920–929. [CrossRef]
53. Metzger, T.; Irawan, A.; Tsotsas, E. Influence of pore structure on drying kinetics: A pore network study. *AIChE J.* **2007**, *53*, 3029–3041. [CrossRef]
54. Al-Futaisi, A.; Patzek, T.W. Extension of Hoshen–Kopelman algorithm to non-lattice environments. *Phys. A Stat. Mech. Appl.* **2003**, *321*, 665–678. [CrossRef]
55. Hoshen, J. On the application of the enhanced Hoshen–Kopelman algorithm for image analysis. *Pattern Recognit. Lett.* **1998**, *19*, 575–584. [CrossRef]
56. Xu, B.; Yortsos, Y.C.; Salin, D. Invasion percolation with viscous forces. *Phys. Rev. E* **1998**, *57*, 739–751. [CrossRef]
57. Lenormand, R. Liquids in porous media. *J. Fluid Mech.* **1990**, *2*, SA79–SA88. [CrossRef]
58. Metzger, T.; Tsotsas, E.; Prat, M. Pore-Network Models: A Powerful Tool to Study Drying at the Pore Level and Understand the Influence of Structure on Drying Kinetics. In *Modern Drying Technology: Tools at Different Scales*; Tsotsas, E., Mujumdar, A.S., Eds.; Wiley: Weinheim, Germany, 2007; pp. 57–102.
59. Lenormand, R.; Touboul, E.; Zarcone, C. Numerical models and experiments on immiscible displacements in porous media. *J. Fluid Mech.* **1988**, *189*, 165–187. [CrossRef]
60. Prat, M.; Bouleux, F. Drying of capillary porous media with a stabilized front in two dimensions. *Phys. Rev. E* **1999**, *60*, 5647–5656. [CrossRef]
61. Vorhauer, N.; Tsotsas, E.; Prat, M. Temperature gradient induced double stabilization of the evaporation front within a drying porous medium. *Phys. Rev. Fluids* **2018**, *3*, 217. [CrossRef]

Article

Extreme Learning Machine-Based Model for Solubility Estimation of Hydrocarbon Gases in Electrolyte Solutions

Narjes Nabipour [1], Amir Mosavi [2,3,4,5], Alireza Baghban [6], Shahaboddin Shamshirband [7,8,*] and Imre Felde [9]

1 Institute of Research and Development, Duy Tan University, Da Nang 550000, Vietnam; narjesnabipour@duytan.edu.vn
2 Kando Kalman Faculty of Electrical Engineering, Obuda University, 1034 Budapest, Hungary; amir.mosavi@kvk.uni-obuda.hu
3 School of Built the Environment, Oxford Brookes University, Oxford OX30BP, UK
4 Faculty of Health, Queensland University of Technology, 130 Victoria Park Road, Brisbane, QLD 4059, Australia
5 Institute of Structural Mechanics, Bauhaus Universität-Weimar, D-99423 Weimar, Germany
6 Chemical Engineering Department, Amirkabir University of Technology, Mahshahr Campus, Mahshahr, Iran; baghban1369@gmail.com
7 Department for Management of Science and Technology Development, Ton Duc Thang University, Ho Chi Minh City, Vietnam
8 Faculty of Information Technology, Ton Duc Thang University, Ho Chi Minh City, Vietnam
9 John von Neumann Faculty of Informatics, Obuda University, 1034 Budapest, Hungary; felde@uni-obuda.hu
* Correspondence: shahaboddin.shamshirband@tdtu.edu.vn

Received: 26 November 2019; Accepted: 7 January 2020; Published: 9 January 2020

Abstract: Calculating hydrocarbon components solubility of natural gases is known as one of the important issues for operational works in petroleum and chemical engineering. In this work, a novel solubility estimation tool has been proposed for hydrocarbon gases—including methane, ethane, propane, and butane—in aqueous electrolyte solutions based on extreme learning machine (ELM) algorithm. Comparing the ELM outputs with a comprehensive real databank which has 1175 solubility points yielded R-squared values of 0.985 and 0.987 for training and testing phases respectively. Furthermore, the visual comparison of estimated and actual hydrocarbon solubility led to confirm the ability of proposed solubility model. Additionally, sensitivity analysis has been employed on the input variables of model to identify their impacts on hydrocarbon solubility. Such a comprehensive and reliable study can help engineers and scientists to successfully determine the important thermodynamic properties, which are key factors in optimizing and designing different industrial units such as refineries and petrochemical plants.

Keywords: hydrocarbon gases; solubility; natural gas; extreme learning machines; electrolyte solution; prediction model; big data; data science; deep learning; chemical process model; machine learning

---

1. Introduction

Solubility of hydrocarbon and nonhydrocarbon gases—i.e., mixtures of methane, ethane, propane, CO_2, and N_2 in aqueous phases—is known as one of the important practical and theoretical challenges in petroleum, geochemical, and chemical engineering. This property has an effective role in different processes, such as achieving optimum conditions for oil and gas transportation, gas hydrate formation, designing thermal separation processes, gas sequestration for protecting environment, and coal gasification. Petroleum reservoirs normally have some natural gases with aqueous solution at high-pressure and high-temperature conditions so that the solubility of gas becomes attractive for engineers [1–8]. In production and transportation of hydrocarbons, it is possible that water content of gas undergoes an alteration in phase from vapor to ice and gas hydrates. The crystalline solid phases called gas hydrates are created when small-sized gas molecules are trapped in lattice of water molecules. Creation of hydrates can cause major flow assurance problems during production and transportation of hydrocarbons steps such as pipeline blockage, corrosion, and many other issues resulted from the two-phase flow [1,9–11].

In the recent years, investigations on CO_2 solubility in aqueous electrolyte solutions have grown significantly as well as they are related to CO_2 capture and storage. It is a clear fact that the dominant cause of global warming is emission of CO_2 gas generated from fossil fuels so its sequestration and disposal in the ocean have been known as a reasonable choice to overcome global warming problems [12–14]. Simulation of enhanced oil recovery, design of supercritical extraction, and optimization of CO_2 dissolution in the ocean need a comprehensive knowledge about carbon dioxide solubility in aqueous electrolytes solutions [13–15].

Investigation of natural gas phase behavior in aqueous solutions in different operational conditions is known one of the important issues in the industry, which has wide applications for avoiding problems in designing and optimization of gas processing. In the literature, there are different solubility datasets for various gas–liquid systems. These datasets mostly include hydrocarbons' dissolution in water/brine systems [1,4,5,9,16–20] and non-hydrocarbons such as CO_2 and N_2 dissolution in water/brine systems [7,12–14,18,21–24]. A brief summary of the hydrocarbon systems datasets is shown in Table 1 for hydrocarbons. The experimental data of water content of hydrocarbons and non-hydrocarbons are limited because of difficulties in measurement of the low water content gases at high pressure and low temperature. Mohammadi and coworkers expressed that an accurate estimation of water content can be obtained by gas solubility data, therefore, they overcame the complexities of experimental determination of the water content in natural gases [1]. Due to limited number of measurement data, wide attempts have been made to model and describe the gas-liquid equilibrium in aqueous electrolyte solutions. There are several thermodynamic models which uses the Henry's constant, activity coefficient, and cubic equations of state to obtain more information about the equilibrium conditions. The changes of Henry's constant for the pressure lower than 5 MPa are negligible and it is dominantly affected by temperatures [19]. The high dependency on temperature is obvious at low temperature and also the nonlinear decreasing relationship is observed at high temperatures [25]. Furthermore, there is just a limited number of Henry's constants for hydrocarbon systems at low temperature. According to this fact, there are several drawbacks in applying the Henry's law, whereas it has great ability for accurate prediction of solubility. As an example, it is suitable for dilute solutions or near-ideal solutions [26]. Additionally, this method is correct for single compounds in no chemical reaction conditions for aqueous phase. Another method is cubic EOS which has several advantages such as small number of parameters, computational efficiency, and ease of performance [3,4,21]. The EOSs were proposed originally for pure fluids, after that, their applications were expanded for mixtures by combining the constants from different pure components. This extension can be done by different methods such as Dalton's law of additive partial pressures and Amagat's rule of additive volumes [5]. For complex compounds, there are some limitations in accuracy of EOS which highlight the importance of empirical adjustments by dealing with the binary interaction parameters. In order to determine these parameters, a reliable source of experimental data for vapor-liquid equilibrium is required which induces some uncertainty into EOSs [7].

Due to above discussions, development of an accurate and reliable approach for estimation of solubility of hydrocarbons and non-hydrocarbons in aqueous electrolyte solutions has been highlighted. Nowadays, machine learning approaches have shown extensive applications in different topics [27–35]. This work organizes a novel artificial intelligence method called extreme learning machine (ELM) to estimate solubility of hydrocarbons in aqueous electrolyte mixtures in terms of types of gas, mole fractions of gases, pressure, temperature, and ionic strength.

Table 1. Details of experimental hydrocarbons solubility in aqueous electrolyte solutions.

Author	P (Mpa)	T (°C)	Composition	Mole Fraction of the Components in the Gaseous Phase
Culberson et al.	0.8–69.61	37.78–171.11	Pure water	C1: 0.0000698–0.0033
Kiepe et al.	0.304–10.23	40–100.14	Pure water, LiBr, KBr, LiCl, KCl	C1: 0.00003–0.00154
Chapoy et al.	0.357–18	1.98–95.01	Pure water	C1: 0.000204–0.002459 C2: 0.0000147–0.0000674 C3: 0.0000321–0.0002694 C4: 0.00000387–0.00001121
Marinakis et al.	6.22–20.1	1.4–25.98	Pure water, NaCl	C1: 0.00099–0.00282 C2: 0.000038–0.000249 C3: 0.000006–0.000042
Crovetto et al.	1.327–6.451	24.35–245.15	Pure water	C1: 0.0002124–0.0010337
Wang et al.	1–40.03	2.5–30.05	Pure water	C1: 0.000563–0.004049 C2: 0.0000986–0.000864
Amirjafari	4.66–56.16	54.44–104.44	Pure water	C1: 0.00045–0.0037 C2: 0.000119–0.001768 C3: 1.9×10^{-5}–0.001863
O'Sullivan et al.	10.2–62	51.5–125	Pure water, NaCl	C1: 0.000805–0.0043 C2: 0.000825–0.001438
Michels et al.	4.09–45.89	25–150	Pure water, NaCl, LiCl, NaBr, NaI, $CaCl_2$	C1: 0.000173–0.00269
Mohammadi et al.	1.14–31.1	4.65–24.75	Pure water	C1: 0.000313–0.00311
Vul'fson et al.	2.53–60.8	19.95–79.95	Pure water	C1: 0.000361–0.004328
Dhima	2.5–100	71	Pure water	C1: 0.000127–0.005085 C2: 0.000821–0.001398 C4: 0.000021–0.000103

2. Methodology

2.1. Experimental Dataset Collection

In order to construct a highly accurate and comprehensive model capable of estimating the solubility of mixtures of hydrocarbons in aqueous electrolyte solutions, a comprehensive databank was provided based on existing experimental data in Table 1. This databank contains total number of 1175 solubility points for hydrocarbons (881 and 294 points for training and testing phases, respectively) (see Table S1 of data set in Supplementary Materials). According to the literature [1,4,5,9,16–20], the solubility of gases in these systems is highly function of aqueous solutions, pressure, temperature, and gaseous phase composition. The aqueous phase composition was change into ionic strength (I) from salt concentrations to reduce dimensions of modeling process. The following equation presents the relationships between ionic strength, valance of charged ions (z_i), and molar concentration of each ion (m_i).

$$I = \frac{1}{2}\sum m_i |z_i|^2 \quad (1)$$

In this study, the solubility of hydrocarbons is predicted in terms of concentration of components in gaseous mixture, ionic strength of solution, temperature, and pressure.

$$\eta_h = f(C1, C2, C3, C4, I, P, T, idx) \quad (2)$$

In which, η_h represents the hydrocarbon solubility in aqueous phase; $C(1–4)$ are known as the methane, ethane, propane, and butane mole fraction in gas phase (0–99.99); I denotes the ionic strength based on molarity (0–37.35); T denotes the temperature in terms of °C (1.4–245.15); P shows the pressure in MPa (0.3–100), and idx symbolizes the index of fraction whose solubility is to be determined (1,2,3,4).

2.2. Extreme Learning Machine

Huang proposed a new intelligence method based on single-layer feedforward neural network (SLFFNN) called extreme learning machine to satisfy the drawbacks of gradient-based algorithms such low training speed and low learning rate. In the ELM algorithm, the hidden nodes are selected randomly and the weights of output of the SLFFNN are calculated by applying Moore–Penrose generalized inverse [36,37].

The scheme of ELM algorithm is demonstrated in Figure 1. By assuming N training sets such as $(x_i, y_i) \in R^n \times R^m$ for L hidden nodes, the SLFFNN algorithms can be written as

$$\sum_{i=1}^{L} \beta_i f_i(x_j) = \sum_{i=1}^{L} \beta_i f_i(a_i.b_i.x) \qquad j = 1, \ldots, N \tag{3}$$

In which, $a_i = [a_{i1}, \ldots, a_{in}]^T$ points to input weights matrix which is related to hidden nodes, $\beta_i = [\beta_{i1}, \ldots, \beta_{im}]^T$ represents the output weights matrix which is related to hidden nodes, and b_i symbolizes the hidden layer bias.

$$\sum_{i=1}^{L} \beta_i f_i(x_j) = H\beta \tag{4}$$

In which, $\beta = [\beta_1, \ldots, \beta_L]$ and $h(x) = [h_1(x), \ldots h_L(x)]$ are known as the hidden layer output matrix and the output weight matrix.

The first step of this model is the random calculation of input weight and the bias of hidden layer for the training phase. Then, for determining these values, the hidden layer matrix is obtained by utilization of input variables. Then, the SLFFNN training is changed to a least-square problem. The ELM algorithms implement regularization theory to define a target function as [38–40]

$$\min L_{ELM} = \frac{1}{2}\|\beta\|^2 + \frac{c}{2}\|T - H\beta\|^2 \tag{5}$$

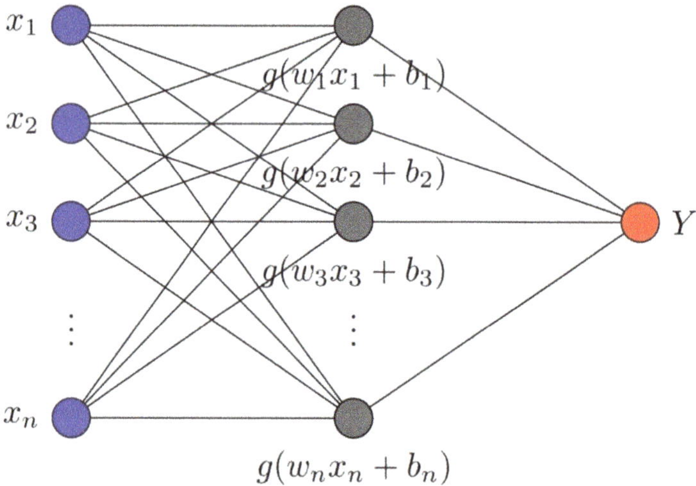

Figure 1. Structure of ELM algorithm.

3. Results and Discussion

In this study, the solubility of hydrocarbons in the aqueous electrolyte phase is determined based on ELM algorithm. To this end, the sigmoid function is set as activation function and the input weights were initialized randomly in range of (−1, 1). Additionally, the number of nodes in the hidden layers was estimated as 30 based on the lowest value of RMSE as determined in Figure 2. As shown, after

30 nodes, by increasing complexity of model, the testing error increased so the optimum structure of the algorithm has 30 nodes to prohibit overfitting.

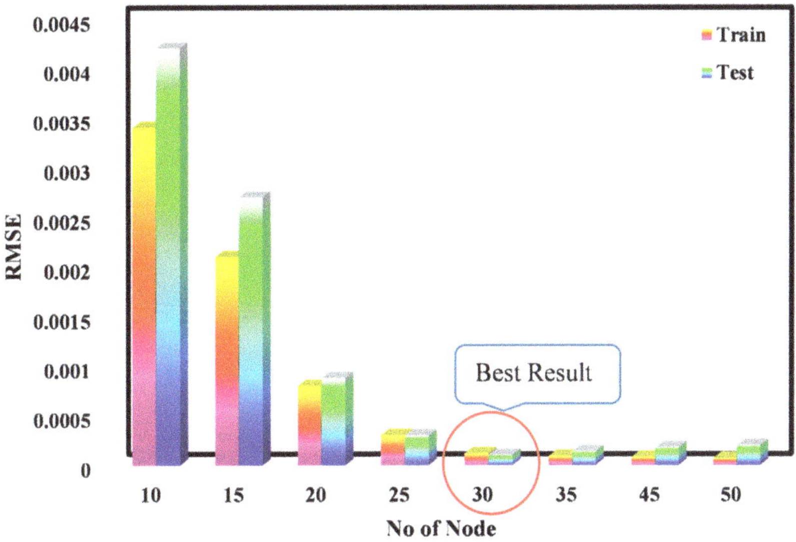

Figure 2. Obtaining optimum structure of proposed algorithm.

In the following, the statistical results of the estimation of hydrocarbon solubility are inserted in Table 2. The following equations are used to achieve this end:

$$\text{Mean relative error}\left(\text{MRE}\right) = \frac{100}{N}\Sigma_{i=1}^{N}\left(\frac{X_i^{actual} - X_i^{predicted}}{X_i^{actual}}\right) \tag{6}$$

$$\text{Root mean square error (RMSE)} = \sqrt{\frac{1}{N}\sum\nolimits_{i-1}^{N}\left(\left(X_i^{actual} - X_i^{predicted}\right)^2\right)} \tag{7}$$

$$\text{Mean squared error (MSE)} = \frac{1}{N}\sum_{i=1}^{N}\left(X_i^{actual} - X_i^{predicted}\right)^2 \tag{8}$$

$$\text{R} - \text{squared (R2)} = 1 - \frac{\Sigma_{i=1}^{N}\left(X_i^{actual} - X_i^{predicted}\right)^2}{\Sigma_{i=1}^{N}\left(X_i^{actual} - \overline{X^{actual}}\right)^2} \tag{9}$$

As shown in Table 2, the MRE, MSE, and RMSE are determined as 22.049, 1.33285×10^{-8}, and 0.0001 for training phase respectively. Moreover, for testing phase, MRE = 22.054, MSE = 1.05351×10^{-8} and RMSE = 0.0001 are calculated. The estimated R^2 values are 0.985, 0.987, and 0.985 for training, testing, and overall datasets respectively. These results give the knowledge about the high degree of accuracy for proposed ELM algorithm.

Table 2. Statistical analyses of developed model.

Dataset	R^2	MRE (%)	MSE	RMSE
Training	0.985	22.049	1.33285×10^{-8}	0.0001
Testing	0.987	22.054	1.05351×10^{-8}	0.0001
Overall	0.985	22.050	1.26295×10^{-8}	0.0001

On the one hand, the comparison between the estimated and real hydrocarbons solubility in aqueous electrolyte solutions are shown in Figure 3. This depiction demonstrates an excellent agreement between estimated and real solubility values. Figure 4 also represents the regression plot of actual hydrocarbons solubility versus estimated one. A light cloud of data near the 45° line expresses the validity and accuracy of ELM algorithm. Additionally, Figure 5 also shows the distribution of relative deviations between forecasted and actual hydrocarbons solubility in aqueous solutions. It can be seen that the ELM outputs deviate slightly from the real solubility and most of relative deviations are near to zero. Furthermore, Figure 6 shows the histograms of relative deviations for training and testing phases. In this demonstration, frequency diagram confirms that most of the error points are close to zero and also cumulative axis express the fact that range of deviation is very limited and the highest slope of the cumulative curve occurred near the zero point.

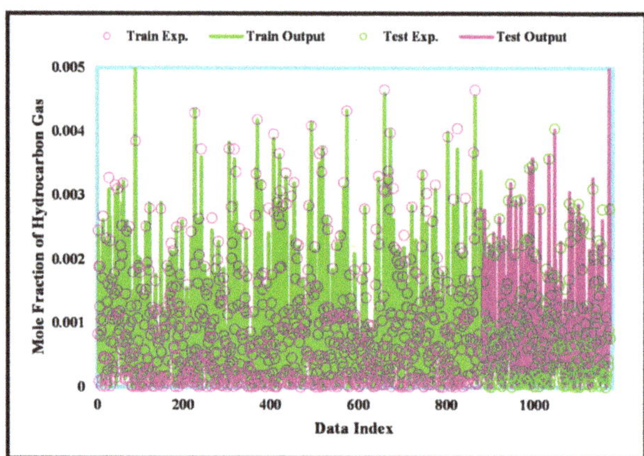

Figure 3. Comparison of actual and estimated solubility of hydrocarbons.

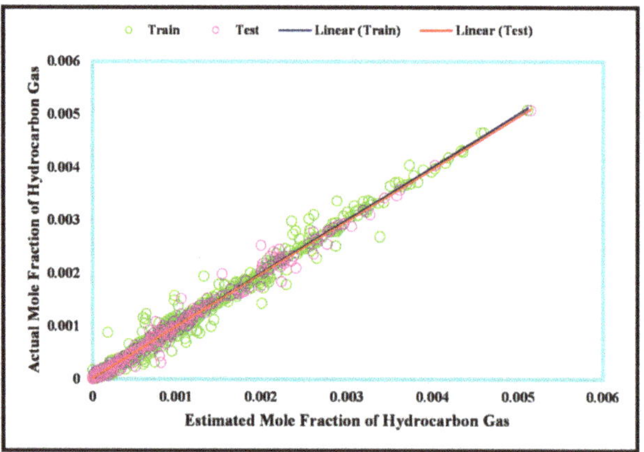

Figure 4. Cross plot of actual and estimated solubility of hydrocarbons.

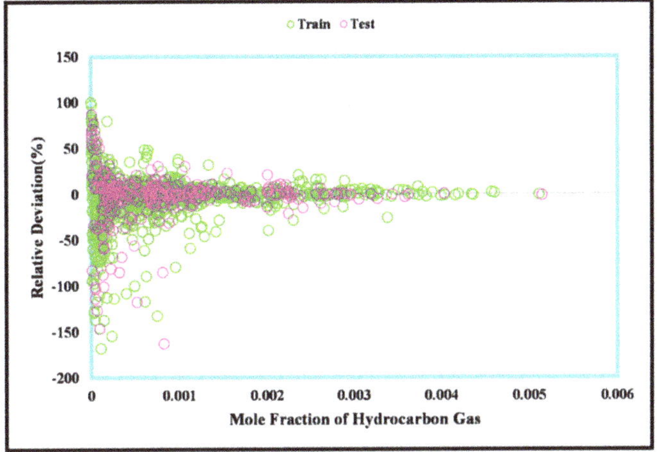

Figure 5. Relative deviation between actual and estimated solubility of hydrocarbons.

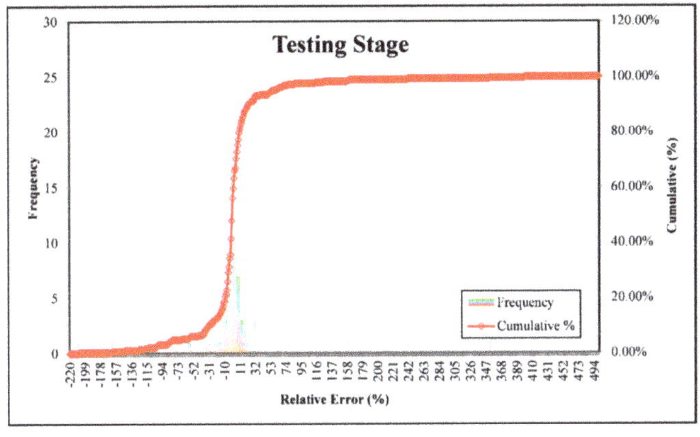

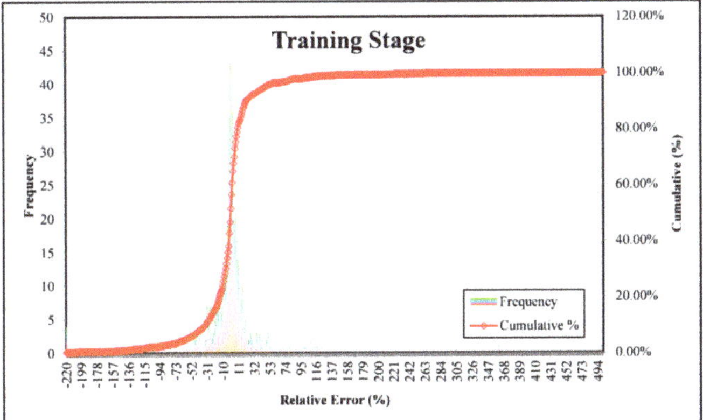

Figure 6. Histogram diagram of relative deviations.

The ELM algorithm implemented in the current work shows an excellent ability in calculation of solubility of hydrocarbons in aqueous phases. One of the important factors which can influence the validation of model is degree of precision of utilized data. In order to clarify the accuracy of solubility databank, the leverage mathematical method is recruited. This method has some rules to identify the suspected solubility data so that a matrix which is known as hat matrix, should be constructed based on formulation [41–45]

$$H = U\left(U^T U\right)^{-1} U^T \tag{10}$$

In which, U symbolizes a matrix of $i \times j$ dimensional. i and j are known as the number of algorithm parameter and training points which are used for determination of critical leverage limit as

$$H^* = 3(j+1)/i \tag{11}$$

In order to detect the reliable zone, there are two standard residual indexes (−3 and 3) which are used in the leverage method. As shown in Figure 7, the reliable area is bound by these two residual indexes and critical leverage limit. The critical straight lines are shown by red and green colors. This plot is known as William's plot. In this plot, normalized residual is depicted versus hat value which is determined from the main diagonal of aforementioned matrix. It is obvious that the major number of solubility data are located in this area which expresses validation of the hydrocarbon solubility databank.

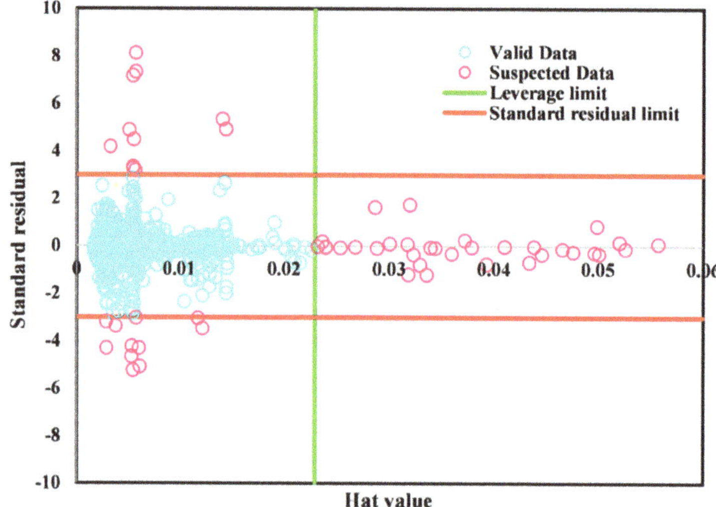

Figure 7. Detection of suspected data for hydrocarbon solubility dataset.

In the most of parametric studies, it is a valuable attempt to identify the effectiveness of all inputs on the target. According to this fact, the sensitivity analysis is employed to investigate effect of concentration of components in gaseous mixture; ionic strength of solution; and temperature and pressure on the solubility of hydrocarbons in aqueous electrolyte systems. To this end, the relevancy factor should be determined as follows for each input parameter [46–54]:

$$r = \frac{\sum_{i=1}^{n} (X_{k,i} - \overline{X_k})\left(Y_i - \overline{Y}\right)}{\sqrt{\sum_{i=1}^{n} (X_{k,i} - \overline{X_k})^2 \sum_{i=1}^{n} (Y_i - \overline{Y})^2}} \tag{12}$$

In which Y_i and $\overline{Y}$ denote the 'i' th output and output average. $X_{k,i}$ and $\overline{X_k}$ are known as 'k' th of input and average of input. Figure 8 shows the relevancy factor for each effective variable of hydrocarbon solubility. It is necessary to explain that the relevancy factor lies in range of −1 to 1 so that the higher absolute value has more impact on hydrocarbon solubility. Furthermore, the positive relevancy factor shows the straight relationship between input and target. The relevancy factors for pressure, temperature, the index of fraction, ionic strength, methane, ethane, propane, and butane mole fraction in gas phase are 0.52, 0.20, −0.48, −0.16, 0.11, 0.06, −0.19, and −0.07 respectively. According to this explanation and results, as pressure, temperature, and mole fraction of methane and ethane increase, the solubility of investigated hydrocarbon increases. Moreover, pressure and mole fraction of ethane in gaseous phase are the most and least effective parameters for determination of solubility of hydrocarbons.

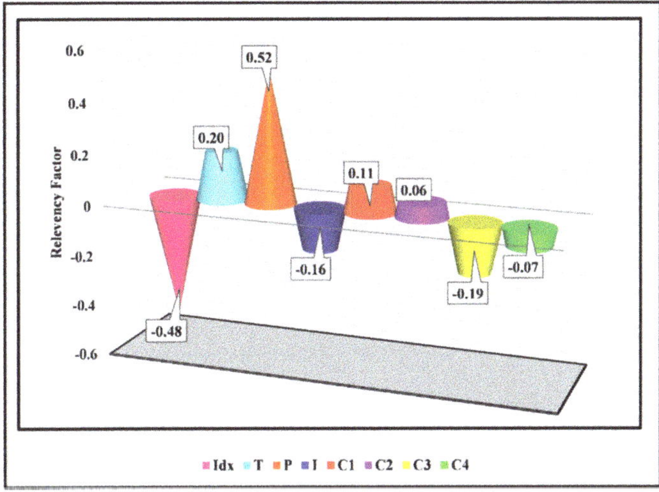

Figure 8. Sensitivity analysis for solubility of hydrocarbons.

4. Conclusions

The hydrocarbon solubility in aqueous electrolyte phases at high temperature and pressure conditions is known as a major effective parameter in variety of applications for petroleum industries and chemical engineering. Numerous attempts have been made in the current study to suggest a highly accurate and comprehensive predicting tool on the basis of extreme learning machine to calculate hydrocarbons solubility in wide ranges of operational conditions. Comparing the ELM outputs with a comprehensive real databank which has 1175 solubility points concluded to R-squared values of 0.985 and 0.987 for training and testing phases respectively. The excellent agreements of ELM and real hydrocarbon solubility values express that the ELM algorithm is a valuable tool for design and optimization of various processes that are related to vapor-liquid equilibrium. Furthermore, this study gives more information about the intensity of each input parameter on solubility of hydrocarbons. Due to the aforementioned results, this work has potential application in commercial software packages such as CMG and ECLIPSE for simulation of fluid flow in porous media.

Supplementary Materials: The following are available online at http://www.mdpi.com/2227-9717/8/1/92/s1, Table S1: data set.

Author Contributions: Conceptualization, A.B. and S.S.; Methodology, N.N., A.M., and A.B.; Software, N.N., A.M., A.B., S.S., and I.F.; Validation, N.N., A.M., A.B., S.S., and I.F.; Formal analysis, N.N., A.M., A.B., S.S., and I.F.; Investigation, N.N., A.M., A.B., S.S., and I.F.; Resources, N.N., A.M., A.B., S.S., and I.F.; Data curation, N.N., A.M., A.B., S.S., and I.F.; Writing—Original draft preparation, N.N., A.M., and A.B.; Writing—Review and editing, N.N.,

A.M., A.B., S.S., and I.F.; Visualization, N.N., A.M., A.B., S.S., and I.F.; Project administration and submission, A.M., Supervision, I.F. All authors have read and agreed to the published version of the manuscript.

Funding: This research received no external funding.

Acknowledgments: We acknowledge the financial support of this work by the Hungarian State and the European Union under the EFOP-3.6.1-16-2016-00010 project.

Conflicts of Interest: The authors declare no conflicts of interest.

References

1. Mohammadi, A.H.; Chapoy, A.; Tohidi, B.; Richon, D. Gas solubility: A key to estimating the water content of natural gases. *Ind. Eng. Chem. Res.* **2006**, *45*, 4825–4829. [CrossRef]
2. Chapoy, A.; Haghighi, H.; Tohidi, B. Development of a Henry's constant correlation and solubility measurements of n-pentane, i-pentane, cyclopentane, n-hexane, and toluene in water. *J. Chem. Thermodyn.* **2008**, *40*, 1030–1037. [CrossRef]
3. Chapoy, A.; Mokraoui, S.; Valtz, A.; Richon, D.; Mohammadi, A.H.; Tohidi, B. Solubility measurement and modeling for the system propane–water from 277.62 to 368.16 K. *Fluid Phase Equilibria* **2004**, *226*, 213–220. [CrossRef]
4. Chapoy, A.; Mohammadi, A.H.; Richon, D.; Tohidi, B. Gas solubility measurement and modeling for methane–water and methane–ethane–n-butane–water systems at low temperature conditions. *Fluid Phase Equilibria* **2004**, *220*, 111–119. [CrossRef]
5. Dhima, A.; de Hemptinne, J.-C.; Moracchini, G. Solubility of light hydrocarbons and their mixtures in pure water under high pressure. *Fluid Phase Equilibria* **1998**, *145*, 129–150. [CrossRef]
6. Kiepe, J.; Horstmann, S.; Fischer, K.; Gmehling, J. Experimental determination and prediction of gas solubility data for methane+ water solutions containing different monovalent electrolytes. *Ind. Eng. Chem. Res.* **2003**, *42*, 5392–5398. [CrossRef]
7. Bamberger, A.; Sieder, G.; Maurer, G. High-pressure (vapor+ liquid) equilibrium in binary mixtures of (carbon dioxide+ water or acetic acid) at temperatures from 313 to 353 K. *J. Supercrit. Fluids* **2000**, *17*, 97–110. [CrossRef]
8. Mohammadi, A.H.; Chapoy, A.; Tohidi, B.; Richon, D. Water content measurement and modeling in the nitrogen+ water system. *J. Chem. Eng. Data* **2005**, *50*, 541–545. [CrossRef]
9. Marinakis, D.; Varotsis, N. Solubility measurements of (methane+ ethane+ propane) mixtures in the aqueous phase with gas hydrates under vapour unsaturated conditions. *J. Chem. Thermodyn.* **2013**, *65*, 100–105. [CrossRef]
10. Kondori, J.; Zendehboudi, S.; Hossain, M.E. A review on simulation of methane production from gas hydrate reservoirs: Molecular dynamics prospective. *J. Pet. Sci. Eng.* **2017**, *159*, 754–772. [CrossRef]
11. Kondori, J.; Zendehboudi, S.; James, L. Evaluation of gas hydrate formation temperature for gas/water/salt/alcohol systems: Utilization of extended UNIQUAC model and PC-SAFT equation of state. *Ind. Eng. Chem. Res.* **2018**, *57*, 13833–13855. [CrossRef]
12. Tong, D.; Trusler, J.M.; Vega-Maza, D. Solubility of CO2 in aqueous solutions of CaCl2 or MgCl2 and in a synthetic formation brine at temperatures up to 423 K and pressures up to 40 MPa. *J. Chem. Eng. Data* **2013**, *58*, 2116–2124. [CrossRef]
13. Teng, H.; Yamasaki, A. Solubility of liquid CO2 in synthetic sea water at temperatures from 278 K to 293 K and pressures from 6.44 MPa to 29.49 MPa, and densities of the corresponding aqueous solutions. *J. Chem. Eng. Data* **1998**, *43*, 2–5. [CrossRef]
14. Lucile, F.; Cézac, P.; Contamine, F.; Serin, J.-P.; Houssin, D.; Arpentinier, P. Solubility of carbon dioxide in water and aqueous solution containing sodium hydroxide at temperatures from (293.15 to 393.15) K and pressure up to 5 MPa: Experimental measurements. *J. Chem. Eng. Data* **2012**, *57*, 784–789. [CrossRef]
15. Nighswander, J.A.; Kalogerakis, N.; Mehrotra, A.K. Solubilities of carbon dioxide in water and 1 wt.% sodium chloride solution at pressures up to 10 MPa and temperatures from 80 to 200.°C. *J. Chem. Eng. Data* **1989**, *34*, 355–360. [CrossRef]
16. Dhima, A.; de Hemptinne, J.-C.; Jose, J. Solubility of hydrocarbons and CO2 mixtures in water under high pressure. *Ind. Eng. Chem. Res.* **1999**, *38*, 3144–3161. [CrossRef]
17. Michels, A.; Gerver, J.; Bijl, A. The influence of pressure on the solubility of gases. *Physica* **1936**, *3*, 797–808. [CrossRef]

18. O'Sullivan, T.D.; Smith, N.O. Solubility and partial molar volume of nitrogen and methane in water and in aqueous sodium chloride from 50 to 125.deg. and 100 to 600 atm. *J. Phys. Chem.* **1970**, *74*, 1460–1466.
19. Vul'fson, A.; Borodin, O. A thermodynamic analysis of the solubility of gases in water at high pressures and supercritical temperatures. *Russ. J. Phys. Chem. A* **2007**, *81*, 510–514. [CrossRef]
20. Wang, L.-K.; Chen, G.-J.; Han, G.-H.; Guo, X.-Q.; Guo, T.-M. Experimental study on the solubility of natural gas components in water with or without hydrate inhibitor. *Fluid Phase Equilibria* **2003**, *207*, 143–154. [CrossRef]
21. Chapoy, A.; Mohammadi, A.H.; Tohidi, B.; Richon, D. Gas solubility measurement and modeling for the nitrogen+ water system from 274.18 K to 363.02 K. *J. Chem. Eng. Data* **2004**, *49*, 1110–1115. [CrossRef]
22. Prutton, C.; Savage, R. The solubility of carbon dioxide in calcium chloride-water solutions at 75, 100, 120 and high pressures1. *J. Am. Chem. Soc.* **1945**, *67*, 1550–1554. [CrossRef]
23. Bando, S.; Takemura, F.; Nishio, M.; Hihara, E.; Akai, M. Solubility of CO2 in aqueous solutions of NaCl at (30 to 60) C and (10 to 20) MPa. *J. Chem. Eng. Data* **2003**, *48*, 576–579. [CrossRef]
24. Smith, N.O.; Kelemen, S.; Nagy, B. Solubility of natural gases in aqueous salt solutions—II: Nitrogen in aqueous NaCl, CaCl2, Na2SO4 and MgSO4 at room temperatures and at pressures below 1000 psia. *Geochim. Cosmochim. Acta* **1962**, *26*, 921–926. [CrossRef]
25. Crovetto, R.; Fernández-Prini, R.; Japas, M.L. Solubilities of inert gases and methane in H2O and in D2O in the temperature range of 300 to 600 K. *J. Chem. Phys.* **1982**, *76*, 1077–1086. [CrossRef]
26. Battino, R.; Clever, H.L. The solubility of gases in liquids. *Chem. Rev.* **1966**, *66*, 395–463. [CrossRef]
27. Kang, X.; Liu, C.; Zeng, S.; Zhao, Z.; Qian, J.; Zhao, Y. Prediction of Henry's law constant of CO2 in ionic liquids based on SEP and Sσ-profile molecular descriptors. *J. Mol. Liq.* **2018**, *262*, 139–147. [CrossRef]
28. Kang, X.; Qian, J.; Deng, J.; Latif, U.; Zhao, Y. Novel molecular descriptors for prediction of H2S solubility in ionic liquids. *J. Mol. Liq.* **2018**, *265*, 756–764. [CrossRef]
29. Kang, X.; Zhao, Y.; Li, J. Predicting refractive index of ionic liquids based on the extreme learning machine (ELM) intelligence algorithm. *J. Mol. Liq.* **2018**, *250*, 44–49. [CrossRef]
30. Qiao, W.; Huang, K.; Azimi, M.; Han, S. A Novel Hybrid Prediction Model for Hourly Gas Consumption in Supply Side Based on Improved Machine Learning Algorithms. *IEEE Access* **2019**, *7*, 88218–88230. [CrossRef]
31. Qiao, W.; Lu, H.; Zhou, G.; Azimi, M.; Yang, Q.; Tian, W. A hybrid algorithm for carbon dioxide emissions forecasting based on improved lion swarm optimizer. *J. Clean. Prod.* **2020**, *244*, 118612. [CrossRef]
32. Hemmati-Sarapardeh, A.; Hajirezaie, S.; Soltanian, M.R.; Mosavi, A.; Nabipour, N.; Shamshirband, S.; Chau, K.W. Modeling natural gas compressibility factor using a hybrid group method of data handling. *Eng. Appl. Comput. Fluid Mech.* **2020**, *14*, 27–37. [CrossRef]
33. Qiao, W.; Yang, Z.; Kang, Z.; Pan, Z. Short-term natural gas consumption prediction based on Volterra adaptive filter and improved whale optimization algorithm. *Eng. Appl. Artif. Intell.* **2020**, *87*, 103323. [CrossRef]
34. Choubin, B.; Abdolshahnejad, M.; Moradi, E.; Querol, X.; Mosavi, A.; Shamshirband, S.; Ghamisi, P. Spatial hazard assessment of the PM10 using machine learning models in Barcelona, Spain. *Sci. Total Environ.* **2020**, *701*, 134474. [CrossRef]
35. Zhao, Y.; Gao, J.; Huang, Y.; Afzal, R.M.; Zhang, X.; Zhang, S. Predicting H 2 S solubility in ionic liquids by the quantitative structure–property relationship method using S σ-profile molecular descriptors. *RSC Adv.* **2016**, *6*, 70405–70413. [CrossRef]
36. Huang, G.-B.; Zhu, Q.-Y.; Siew, C.-K. Extreme learning machine: A new learning scheme of feedforward neural networks. *Neural Netw.* **2004**, *2*, 985–990.
37. Huang, G.-B.; Zhu, Q.-Y.; Siew, C.-K. Extreme learning machine: Theory and applications. *Neurocomputing* **2006**, *70*, 489–501. [CrossRef]
38. Bengio, Y. Learning deep architectures for AI. *Found. Trends®Mach. Learn.* **2009**, *2*, 1–127. [CrossRef]
39. Liu, Q.; Yin, J.; Leung, V.C.; Zhai, J.-H.; Cai, Z.; Lin, J. Applying a new localized generalization error model to design neural networks trained with extreme learning machine. *Neural Comput. Appl.* **2016**, *27*, 59–66. [CrossRef]
40. Rao, C.R.; Mitra, S.K. Further contributions to the theory of generalized inverse of matrices and its applications. *Sankhyā Indian J. Stat. Ser. A* **1971**, *33*, 289–300.
41. Bemani, A.; Baghban, A.; Mohammadi, A.H. An insight into the modeling of sulfur solubility of sour gases in supercritical region. *J. Pet. Sci. Eng.* **2019**, *184*, 106459. [CrossRef]

42. Razavi, R.; Bemani, A.; Baghban, A.; Mohammadi, A.H.; Habibzadeh, S. An insight into the estimation of fatty acid methyl ester based biodiesel properties using a LSSVM model. *Fuel* **2019**, *243*, 133–141. [CrossRef]
43. Mesbah, M.; Soroush, E.; Azari, V.; Lee, M.; Bahadori, A.; Habibnia, S. Vapor liquid equilibrium prediction of carbon dioxide and hydrocarbon systems using LSSVM algorithm. *J. Supercrit. Fluids* **2015**, *97*, 256–267. [CrossRef]
44. Rousseeuw, P.J.; Leroy, A.M. *Robust Regression and Outlier Detection*; John Wiley & Sons: Hoboken, NJ, USA, 2005.
45. Razavi, R.; Sabaghmoghadam, A.; Bemani, A.; Baghban, A.; Chau, K.-W.; Salwana, E. Application of ANFIS and LSSVM strategies for estimating thermal conductivity enhancement of metal and metal oxide based nanofluids. *Eng. Appl. Comput. Fluid Mech.* **2019**, *13*, 560–578. [CrossRef]
46. Baghban, A.; Kahani, M.; Nazari, M.A.; Ahmadi, M.H.; Yan, W.-M. Sensitivity analysis and application of machine learning methods to predict the heat transfer performance of CNT/water nanofluid flows through coils. *Int. J. Heat Mass Transf.* **2019**, *128*, 825–835. [CrossRef]
47. Baghban, A.; Kardani, M.N.; Mohammadi, A.H. Improved estimation of Cetane number of fatty acid methyl esters (FAMEs) based biodiesels using TLBO-NN and PSO-NN models. *Fuel* **2018**, *232*, 620–631. [CrossRef]
48. Baghban, A.; Adelizadeh, M. On the determination of cetane number of hydrocarbons and oxygenates using Adaptive Neuro Fuzzy Inference System optimized with evolutionary algorithms. *Fuel* **2018**, *230*, 344–354. [CrossRef]
49. Zarei, F.; Rahimi, M.R.; Razavi, R.; Baghban, A. Insight into the experimental and modeling study of process intensification for post-combustion CO2 capture by rotating packed bed. *J. Clean. Prod.* **2019**, *211*, 953–961. [CrossRef]
50. Baghban, A.; Pourfayaz, F.; Ahmadi, M.H.; Kasaeian, A.; Pourkiaei, S.M.; Lorenzini, G. Connectionist intelligent model estimates of convective heat transfer coefficient of nanofluids in circular cross-sectional channels. *J. Therm. Anal. Calorim.* **2018**, *132*, 1213–1239. [CrossRef]
51. Bemani, A.; Baghban, A.; Shamshirband, S.; Mosavi, A.; Csiba, P.; Varkonyi-Koczy, A.R. Applying ANN, ANFIS, and LSSVM Models for Estimation of Acid Solvent Solubility in Supercritical CO2. *Preprints* 2019. [CrossRef]
52. Shamshirband, S.; Hadipoor, M.; Baghban, A.; Mosavi, A.; Bukor, J.; Várkonyi-Kóczy, A.R. Developing an ANFIS-PSO Model to Predict Mercury Emissions in Combustion Flue Gases. *Mathematics* **2019**, *7*, 965. [CrossRef]
53. Shabani, S.; Samadianfard, S.; Sattari, M.T.; Mosavi, A.; Shamshirband, S.; Kmet, T.; Várkonyi-Kóczy, A.R. Modeling Pan Evaporation Using Gaussian Process Regression K-Nearest Neighbors Random Forest and Support Vector Machines; Comparative Analysis. *Atmosphere* **2020**, *11*, 66. [CrossRef]
54. Ouaer, H.; Hosseini, A.H.; Nait Amar, M.; El Amine Ben Seghier, M.; Ghriga, M.A.; Nabipour, N.; Andersen, P.Ø.; Mosavi, A.; Shamshirband, S. Rigorous Connectionist Models to Predict Carbon Dioxide Solubility in Various Ionic Liquids. *Appl. Sci.* **2020**, *10*, 304. [CrossRef]

Article

Acid-Base Flow Battery, Based on Reverse Electrodialysis with Bi-Polar Membranes: Stack Experiments

Jiabing Xia, Gerhart Eigenberger, Heinrich Strathmann and Ulrich Nieken *

Institute of Chemical Process Engineering, University of Stuttgart, Boeblinger Strasse 78, D-70199 Stuttgart, Germany; icvt@icvt.uni-stuttgart.de (J.X.); gerhart.eigenberger@icvt.uni-stuttgart.de (G.E.); heinrich.strathmann@icvt.uni-stuttgart.de (H.S.)

* Correspondence: ulrich.nieken@icvt.uni-stuttgart.de

Received: 19 November 2019; Accepted: 9 January 2020; Published: 11 January 2020

Abstract: Neutralization of acid and base to produce electricity in the process of reverse electrodialysis with bipolar membranes (REDBP) presents an interesting but until now fairly overlooked flow battery concept. Previously, we presented single-cell experiments, which explain the principle and discuss the potential of this process. In this contribution, we discuss experiments with REDBP stacks at lab scale, consisting of 5 to 20 repeating cell units. They demonstrate that the single-cell results can be extrapolated to respective stacks, although additional losses have to be considered. As in other flow battery stacks, losses by shunt currents through the parallel electrolyte feed/exit lines increases with the number of connected cell units, whereas the relative importance of electrode losses decreases with increasing cell number. Experimental results are presented with 1 mole L^{-1} acid (HCl) and base (NaOH) for open circuit as well as for charge and discharge with up to 18 mA/cm^2 current density. Measures to further increase the efficiency of this novel flow battery concept are discussed.

Keywords: electrical energy storage; acid-base neutralization flow battery; reverse electrodialysis with bipolar membranes; stack test results

1. Introduction

The efficient storage and recovery of electrical energy is a key issue in the transition of electric energy generation from fossil fuels towards sustainable sources like solar and wind power. Flow batteries present an attractive solution, in particular for stationary and decentralized applications, since the amount of energy stored is separated from energy conversion. Presently, vanadium redox flow batteries are the most important and best-studied type of flow batteries. In redox flow batteries, the electric potential is generated by the electrochemical reactions at the electrodes.

As an alternative, the reverse electrodialysis (RED) has been proposed, where an electrical potential is generated across ion exchange membranes, which separate salt solutions of different ionic concentrations. In most cases studied, seawater and river water have been used as the two solutions [1]. Such a RED-stack consists of a sequence of cells containing one compartment for seawater and one for river water, separated by cation and anion exchange membranes. However, the open voltage of such a cell is only about 0.15 V, which requires a large number of subsequent cells to obtain a reasonable stack voltage.

An interesting alternative RED-concept, which will be considered in the following, is based on the neutralization reaction of acid and base at the internal interface of a bipolar membrane, the reverse electrodialysis with bipolar membranes (REDBP).

In [2] we presented an overview of previous publications on REDBP [3–5], together with our experimental result for a single REDBP cell unit, showing the potential advantages and discussing the present shortcomings of this—so far fairly overlooked—concept.

In the present contribution we extend our experimental results to REDBP stacks, consisting of 5 to 20 cell units. Figure 1 from [2] is a sketch of the stack under charge (lower half) and discharge conditions (upper half). Only one of several repeating cell units between the electrode chambers is shown and only the required main ionic fluxes and the electrode reactions are depicted.

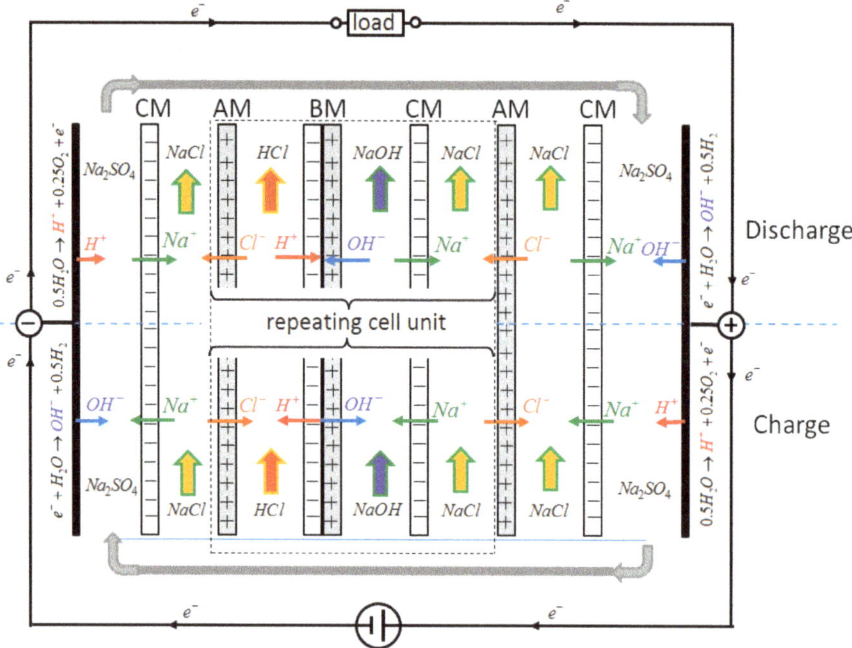

Figure 1. Schematic of main ionic fluxes and electrode reactions of the REDBP (reverse electrodialysis with bipolar membranes) stack used, in the lower part during charge and in the upper part during discharge. Only one of several repeating cell units is shown between the electrode compartments. The respective electrode reactions during charge and discharge are indicated at the electrodes. CM: cation exchange membrane, AM: anion exchange membrane, BM: bipolar membrane.

Each repeating cell unit consists of an acid (HCl) and a base chamber (NaOH) at both sides of the bipolar membrane (BM). A salt compartment (NaCl), separated from the acid or base chamber by an anion exchange membrane (AM) or a cation exchange membrane (CM), provides or accepts the respective Cl^- and Na^+ ions to ensure electroneutrality. To form a REDBP stack, several repeating cell units have to be assembled, with an electrode compartment at each end of the stack.

The main element of each repeating cell unit is the bipolar membrane. Under charge (lower half of Figure 1), a sufficiently high electrical potential has to be applied over the electrodes such that water is split inside the bipolar membrane into H^+ and OH^-. H^+ and OH^- combine with Cl^- and Na^+ from the adjacent salt chambers to produce HCl and NaOH, which can be stored in separate storage tanks. This corresponds to the well-established process of acid and base generation from salt solutions by electrodialysis with bipolar membranes [6]. The respective electrode reactions are water splitting under generation of OH^- and of H_2 at the negative electrode and with H^+ and O_2 generation at the positive electrode. The applied DC power supply transports electrons e^- from the positive to the negative electrode.

If the DC power supply is removed ("open circuit conditions") a potential will establish across each of the bipolar membranes in the stack. It prevents the further recombination of H^+ and OH^- inside each bipolar membrane. The sum of all resulting membrane potentials can be measured between the electrodes of the stack as open circuit voltage (OCV).

If the electrodes are connected via an external load (upper part of Figure 1) the unit can be used as a battery. Now H^+ from HCl and OH^- from NaOH recombine to water inside each bipolar membrane and generate electric energy; the ionic fluxes in each repeating cell unit reverse. The resulting electric potential across all repeating cell units drives the depicted electrode reactions. The electrode reactions transform the ionic into electronic current, which flows through the load of the external circuit. The reaction details inside the bipolar membrane during charge and discharge have been discussed in [2]. They depend on the permselectivities and on the fixed charge densities of the ion exchange membranes used.

Commercially available membranes currently limit acid and base concentrations to 2 M, but long-term stability allows only for operation up to 1 M [2]. Using 1 M acid and base, the theoretical power density is 11 kWh/m^3. Energy densities of vanadium redox flow batteries are in the order of 25 kWh/m^3 [7].

2. Experimental

In our stack experiments hydrochloric acid (HCl) and sodium hydroxide solutions (NaOH)of different concentrations have been used as the respective acid and base, with 0.5 mole/L NaCl as the resulting salt solution. As shown in Figure 2, the respective solutions were circulated through the corresponding compartments from external storage tanks. Through the electrode compartments, a 0.25 mole/L sodium sulphate solution ($Na_2SO_4$was circulated instead of NaCl, to avoid chlorine formation in the electrode reactions. In the set-up of Figure 1, tested in this contribution, an additional salt compartment (NaCl) was placed between the sequence of repeating cell units and each electrode compartment, separated from the electrode compartments by a cation exchange membrane (CM). This should prevent the penetration of Cl^- into the electrode compartments, which would lead to a release of Cl_2 in the electrode reactions. Prime advantages of the chosen electrolytes are their low price, ample availability and low ecological concern.

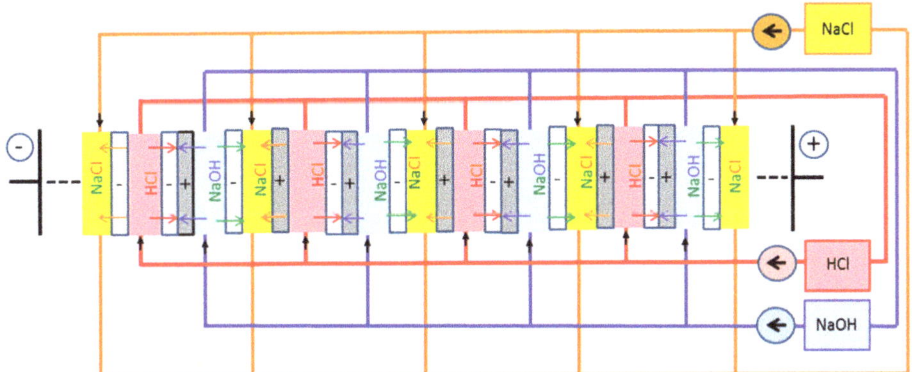

Figure 2. Parallel flow concept for acid, base and salt solution through four repeating cell units in the middle of a stack. The black arrows indicate the flow directions of acid (HCl), of base (NaOH) and of salt solution (NaCl), and the colored arrows indicate the main fluxes of specific ions during discharge, also at OCV.

Table 1 shows the expected electrode reactions during charge and discharge. Their standard potentials represent a voltage sink of 1.23 + 0.83 = 2.06 V, both at charge and at discharge. This loss could be reduced by modified electrode reactions with gas-consuming electrodes, but this was not in the scope of the present study.

Table 1. Electrode reactions during charge and discharge.

		Electrochemical Reaction	Standard Potential
Discharge	Left electrode (negative): Anode	$\frac{1}{2}H_2O \rightarrow H^+ + \frac{1}{4}O_2 \uparrow + e^-$	−1.229 V
	Right electrode (positive): Cathode	$e^- + H_2O \rightarrow OH^- + \frac{1}{2}H_2 \uparrow$	+0.8277 V
Charge	Left electrode (negative): Cathode	$e^- + H_2O \rightarrow OH^- + \frac{1}{2}H_2 \uparrow$	−0.8277 V
	Right electrode (positive): Anode	$\frac{1}{2}H_2O \rightarrow H^+ + \frac{1}{4}O_2 \uparrow + e^-$	+1.229 V

Since the open circuit voltage (OCV) obtained in the single cell experiments with 1 M acid and base was 0.76 V, only a stack with three and more cell units would be able to compensate the electrode losses and deliver a net voltage during discharge. The influence of the electrode losses obviously decreases with an increasing number of cell units in the stack. However, the stack voltage is not fully proportional to the number of cell units. This is partly due to the so-called shunt currents, which flow through the feed and exit lines of the different electrolytes, transforming electrical energy into heat.

2.1. Parallel Flow Concept

In our set-up, a parallel flow concept for the electrolytes has been used, as shown schematically in Figure 2 for four repeating cell units in the middle of a stack. The parallel flow concept has the advantage of reduced pressure drop, compared to a serial flow concept where all of the acid, base and salt solution flows consecutively through adjacent cell units. The potential disadvantage is that the flow through adjacent parallel chambers may differ, if the flow resistances of the chambers are not equal. In addition, a parallel flow concept leads to larger ionic shunt currents through the feed/exit lines, as discussed in the next subsection. The colored arrows in Figure 2 indicate the main ionic fluxes across the membranes at discharge. At a smaller extent, similar fluxes also occur at OCV conditions. They are caused by a breakthrough of co-ions (Cl^- through CM and H^+ through AM), as discussed more detailed in [8]. This breakthrough is responsible for the fact that the open current voltage (OCV) across a repeating cell unit for 1 mol/L HCl and NaOH at 25 °C decreases from the theoretical value of 0.828 V to a measured value of 0.764 V.

Like conventional electrodialysis stacks, the stack is formed from cell frames with pathways for the respective electrolytes, grid spacers in the cell frame windows, with the respective membranes and with rubber gaskets between cell frames and membranes. The cell frames for the stack experiments of 0.5 mm thickness have been obtained from DEUKUM Co, Frickenhausen, Germany (www.deukum.de), see Figure 3. They proved to be well suited for our lab-scale experiments, since they allow feeding of four different electrolytes through the holes in the frame. Their active membrane area is about 100 cm^2. As for the single cell tests, the membranes (fumasep® FKB, fumasep® FAB and fumasep® FBM,) have been provided by Fumatech Co., Bietigheim-Bissingen, Germany. Their properties are listed in Table 2.

Table 2. Properties of membranes fumasep® FAB, fumasep® FBM with fumasep® PEEK as reinforcement, provided by Fumatech [9].

	Type	Reinforcement	Thickness [μm]	IEC [$meqg^{-1}$]	Selectivity [%]	Specific Area Resistance [Ω/cm^2]
FAB	anion	PEEK	100–130	1.0–1.1.	94–97	4–7
FKB	cation	PEEK	100–130	1.2–1.3	98–99	4–6
FBM	bipolar	PEEK	100–130	-	-	

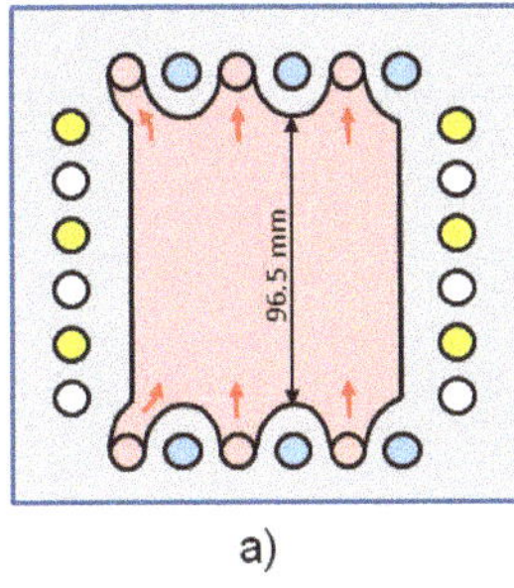

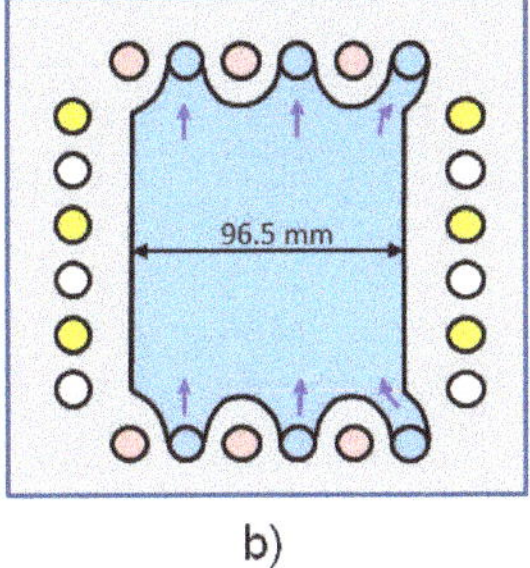

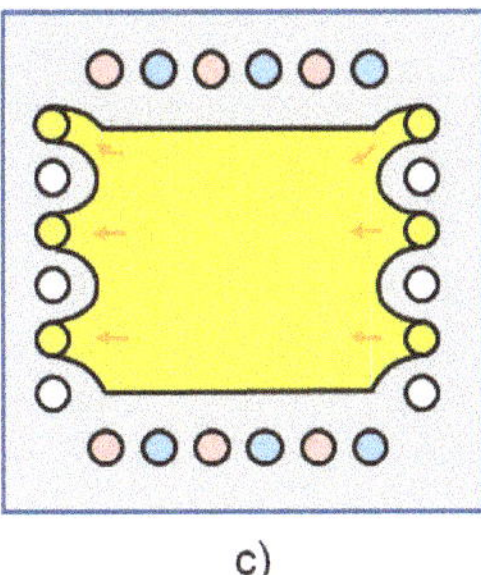

Figure 3. Subsequent cell frames for one repeating cell unit: (**a**) for up-flow of HCl, (**b**) for up-flow of NaOH and (**c**) for cross-flow of the NaCl solution.

Figure 3 shows the flow through subsequent cell frames, (a) for up-flow of HCl, (b) for up-flow of NaOH and (c) for cross-flow of the NaCl solution. Together with the respective membranes, separated by rubber gaskets and grid spacers, the three frames form the repeating cell units shown schematically in Figure 1. The feed and exit flows of the respective electrolytes pass through the holes in the cell frames as indicated by the respective colors in Figure 3. All electrolytes are circulated by separate pumps from reservoirs through the respective compartments as indicated in Figure 2. Further details of the design are given in [8].

2.2. Voltage Measurements

The most direct information during stack operation can be obtained from voltage measurements along the stack. In the single repeating cell unit studied in [2], Haber-Luggin capillaries with calomel electrodes have been used to measure the voltage drop across the respective membranes. To reduce the ample space required for the Haber-Luggin capillaries, in the present stack experiments the voltages between different repeating cell units have been measured by thin Pt wires. The insulated Pt wires were introduced and sealed between two rubber gaskets. To minimize concentration effects, the Pt wires were only positioned in the acid compartments, where ionic conductivities are high and about equal throughout the stack.

In the following, voltage measurements will be presented where Pt wire probes were positioned in the center of the 1st, the 2nd, the 6th, the 11th and the 19th acid cell of the stack. This allows measurement of the voltages across the 1st repeating cell unit, across the first five-cell unit (next to the electrode) and across the second five-cell unit (in the middle of the stack). In order to determine the total voltage of the 20-cell stack, the measured first cell voltage was added to the voltage measured between the first and the 19th acid compartment, assuming that the voltage across the repeating cell units next to both electrodes was equal.

2.3. Operating Conditions

In all cases, Na_2SO_4 electrolyte of 0.25 M was circulated through the electrode chambers and 0.5 M NaCl electrolyte was circulated through the salt chambers, while the concentration of HCl and NaOH, circulated through their chambers, was varied between 0.5 M and 1 M. If not specified differently, the empty-space flow velocity through the electrolyte chambers was adjusted to a mean value of about 2 to 3 cm/s.

3. Results and Discussion

In this section, the measured voltages in stacks consisting of 5 to 20 repeating cell units are presented and discussed for different acid and base concentrations, for different charge and discharge currents and for different charge/discharge periods.

3.1. Open Circuit Voltage (OCV)

With the Pt wire probes, the open circuit voltage can be measured between different cell units as mentioned above, but OCV for the whole stack can also be measured directly between the electrodes of the stack. Table 3 presents OCV results for increasing stack size between 1 and 20 repeating cell units if the electrolytes flow through the stack continuously (constant ion concentrations). In an ideal stack, OCV should amount to the theoretical single cell voltage (0.828 V for 1 M acid and base at 25 °C) times the number of repeating cell units. With the measured single cell voltage of 0.764 V, the theoretical values of the stack are given in the second row. The actually obtained OCV values, measured across the electrodes, are given in the third row. While the measured values are still close to the calculated values for up to 10 repeating cell units, an increasing voltage loss is observed if the cell unit number is raised to 15 and further to 20 cells. Reasons for this trend are discussed in the next section.

Table 3. Calculated and measured open circuit voltage (OCV) for 1 M acid and base at 25 °C, depending on the number of repeating cell units.

Number of Repeating Cell Units	1	5	10	15	20
OCV calculated from single cell voltage	0.764 V	3.82 V	7.64 V	11.46 V	15.28 V
OCV measured across electrodes	0.764 V	3.8 V	7.6 V	11.2 V	14.4 V

3.2. Self-Discharge at OCV

The experimental results of Table 2 have been obtained with stacks where the required electrolytes were continuously pumped through the stack, compensating the effect of any side reactions or shortcuts, which may consume acid and base at OCV. In the OCV experiments reported in Figure 4, the electrolyte flows have been stopped at the beginning of the experiment. Now the resulting drop of the voltage in stacks with different numbers of repeating cell units can be analyzed over time. This should provide indications as to why OCV across stacks with a larger number of repeating cell units increases less than proportional to the number of units, as shown in Table 2. A self-discharge due to limited membrane permselectivities, as indicated by the ionic flux arrows at OCV in Figure 2 and discussed in detail in [2], could contribute to the effect. However, it should not be responsible for the less than proportional increase of OCV with the number of repeating cell units.

Figure 4a–d show the results of the self-discharging tests of (a) 5-cell, (b) 10-cell, (c) 15-cell and (d) 20-cell stacks with initial 1 M acid and base. In a 5-cell stack (Figure 4a), the self-discharge is very low, and OCV remains close to 4V during the whole measurement period. In a 10-cell stack (Figure 4b), the stack voltage is about doubled and a slow self-discharge is followed by an accelerated discharge after about 30 min down to a level of 4 V. If the cell number is further increased (Figure 4c,d), the accelerated drop from the higher stack voltage down to about 4 V occurs at about 20 min for the 15-cell stack, and at about 12 min for the 20-cell stack. This faster decrease of OCV is consistent with the assumption that ionic shunt current through the feed/exit lines of the electrolytes should be responsible for the OCV decrease. A similar influence of shunt currents through parallel feed/exit lines of the electrolytes is well known from redox flow batteries [10–12].

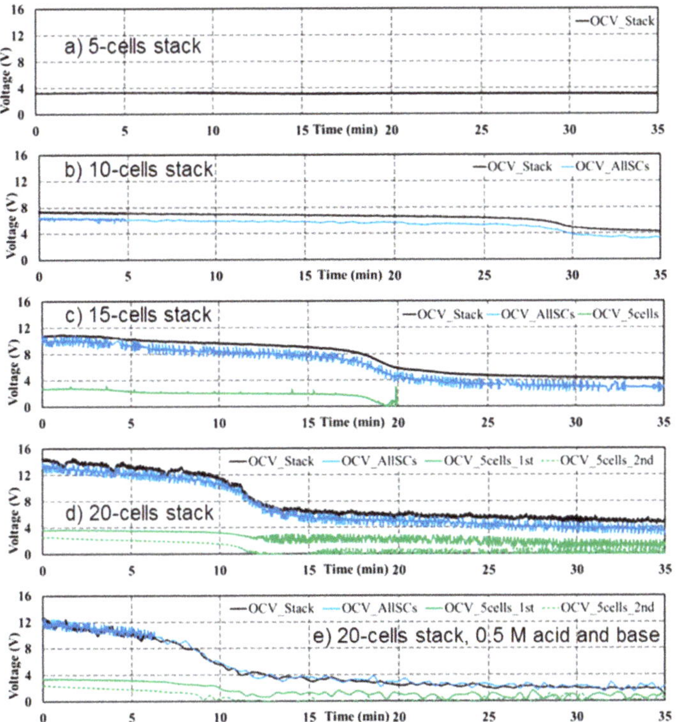

Figure 4. Self-discharge test at 25 °C for different stack sizes and different acid and base concentrations. (**a**) 5-cell stack, 1M acid and base, (**b**) 10-cell stack, 1M acid and base, (**c**) 15-cell stack, 1M acid and base, (**d**) 20-cell stack, 1M acid and base, (**e**) 20-cell stack with 0.5 M acid/base concentration. OCV_Stack (black) has been measured between the electrodes, OCV_AllSCs (blue) between the first and the last repeating cell unit of the stack, OCV_5cells_1st (green solid) and OCV_5cells_2nd (green, dotted) have been measured across the first and second 5-cell unit.

In stacks with 10 and more cells, OCV shows an accelerated decline to about 4 V after a certain period of low decline, which becomes shorter with increasing cell number. An explanation can be drawn from Figure 4d. Here, the difference between OCV over the first 5 cells of the stack (OCV_5cells_1st, green solid line) and OCV over the second 5-cell package (OCV_5cells_2nd, green dotted line) is shown. The voltage over the 5 middle cells was smaller than that over the first 5 cells from the very beginning. It dropped to (almost) zero at the time of the accelerated decline of the total cell voltage, whereas the voltage over the first 5 cells was less affected. The voltage decline indicates that the acid and base have been completely consumed (neutralized) in the middle of the stack, but are still present in the cells next to the electrodes. This can be explained by the ionic shunt currents through the feed/exit lines of the electrolytes. Since H^+ and OH^- have much higher ionic mobility than the other ions present, only the influence of H^+ and OH^- shunt currents in the acid and base feed/exit lines will be considered in the following, using the simplified picture in Figure 5.

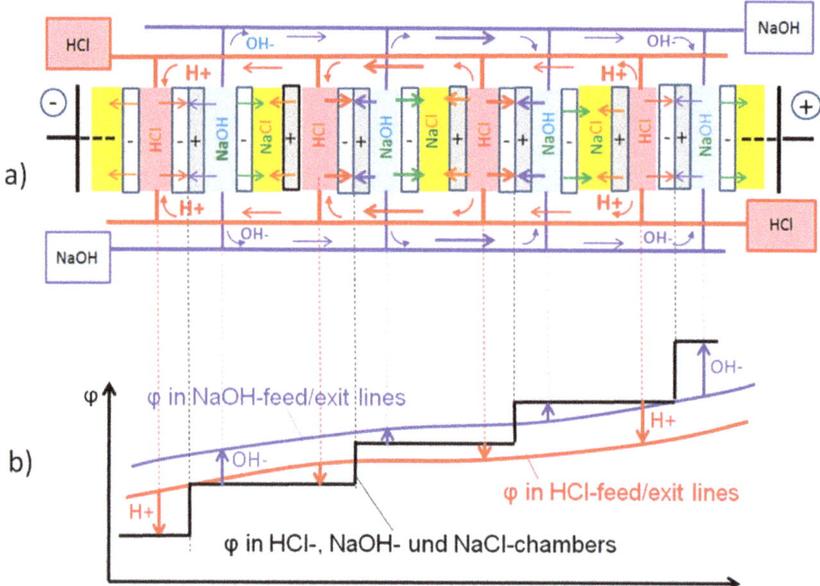

Figure 5. (**a**) Simplified sketch of ionic shunt currents of H^+ (red arrows) and OH^- (blue arrows) through the feed/exit lines of acid and base, and compensating ionic currents inside of four repeating cell units in the middle of the stack. The different arrow thickness indicates the increased shunt currents in the stack center. (**b**) Corresponding electric potential profiles φ across the repeating cell units (black) and in the acid (red) and the base feed/exit lines (blue). The arrows indicate the direction of the ion fluxes between cell compartments and feed/exit lines.

In Figure 5a only four repeating cell units in the middle of the stack are shown, together with their feed/exit lines for acid (red) and base (blue). At OCV, a voltage profile φ (black) develops across the stack as shown in Figure 5b. It mainly consists of the voltage steps in each of the bipolar membranes. The voltage gradient across the stack causes a counter-movement of H^+ ions and of OH^- ions along their feed/exit lines as indicated by the red and blue arrows in Figure 5a. The ions originate from the acid/base compartments at one side of the stack and turn back into the respective compartments at the opposite side. This leads to an ion flux maximum in the stack center. To ensure electroneutrality, the ion fluxes through the feed/exit lines have to be compensated by fluxes of opposite direction through the cell compartments. This leads to the indicated fluxes with neutralization of H^+ and OH^- inside the bipolar membranes, which is maximal in the middle of the stack. Comparable to a simple mixing of acid and base, this neutralization energy is converted into heat, representing a self-discharge with loss of electrical energy. The fluxes of H^+ and OH^- into the bipolar membranes have to be compensated by equivalent fluxes of Cl^- and Na^+ into the adjacent salt compartments.

Arrows in Figure 5b qualitatively depict the driving potentials between the feed/exit lines and the potentials on both sides of the respective bipolar membranes. This explains why the ion fluxes of H^+ and OH^- between the stack cells and the feed/exit lines change direction in the middle of the stack, as indicated by the arrows in Figure 5a,b. The fluxes of H^+ and OH^- through the feed/exit lines accumulate in the stack center as indicated by the arrow thickness. The compensating fluxes inside the repeating cell units are therefore also strongest in the middle of the stack. This leads to the pronounced self-discharge and the resulting drop of OCV across the bipolar membranes in the middle of the stack.

In summary, the ionic shunt currents of H^+ and OH^- ions in the acid/base feed exit lines are driven by the potential along the acid/base feed/exit lines and result in a neutralization reaction in the bipolar membranes. Estimation shows that in the stack used, shunt currents due to H^+ migration through the feed/exit lines of HCl amount to more than 60% and due to OH^- migration through the feed/exit lines, NaOH amounts to more than 30% of the total shunt current [8]. This leads to the consumption of acid and base, primarily in the stack center, which reduces the respective cell voltages. If under OCV the flow of acid and base is stopped, it leads to the gradual neutralization of all acid and base in the stack center as can be concluded from the voltage drop to zero of "OCV_5cells_2nd" after about 12 min in Figure 4d.

In Figure 4e, results are presented where the initial acid/base concentration has been reduced from 1 to 0.5 M. This reduction of 50% should result in an accelerated discharge in about half the time. However, since the reduced concentrations also lead to reduced ionic conductivities and hence to a reduced shunt current, only a small difference between Figure 4d,e can be observed. Again, OCV across the second 5-cell unit in the stack center drops to zero, now after around 10 min.

The above details of the observed OCV decline after stopping the acid/base feed help to elucidate the influence of shunt currents. However, it should be mentioned that shunt currents are only of importance for the REDBP battery during operation. The consumption of acid and base by self-discharge at OCV is limited to the amount present in the stack and does not affect the concentration of acid and base in the storage tanks. This is a general advantage of flow batteries.

To determine all the details of ionic flux interactions, a numerical simulation of the whole stack would be required, where ionic species balances for all ionic species as well as electroneutrality have to be enforced in the whole stack. But this is out of scope for the present contribution.

3.3. Charge and Discharge Behavior

Figure 6 shows the charge and discharge behavior of a 20-cell stack with 1 M acid/base concentration at 25 °C. Starting with OCV conditions, after 1 min a charge current density of 9 mA/cm^2 is imposed for 5 min, followed by zero current for 1 min and a discharge current density of 9 mA/cm^2 for another 5 min. The voltage measured across the stack (black line) is higher under charging and lower under discharging than the voltage measured with the Pt wire probes between all repeating cell units, because the electrode losses need to be added to the blue curve under charge and subtracted under discharge. The mean voltages measured along the 5 min charge and discharge periods are constant. Under constant acid and base concentrations, the measured voltages even remain constant during extended charge and discharge periods, indicating a stable operation. Again, the voltage measured across the second 5-cell package (V5cells_2nd, green dotted line) is lower than V5cells_1st (green solid line) at OCV as well as under charge/discharge, indicating the increased influence of the shunt currents in the stack center.

The absolute difference between the blue and the black curves should, however, be about equal for charge and discharge and vanish at OCV. This would be the case if the blue lines were shifted upwards by 0.7 V, leading to an absolute difference between both curves at charge and discharge of about 3.4 V. This seems to be a reasonable estimate of the losses at the electrodes and across the electrode compartments during charge/discharge with 9 mA/cm^2, since under ideal conditions electrode losses of 2.07 V have been predicted by Table 1. The required voltage shift of 0.7 V points to a measurement error of the Pt wire probes (blue curves). It is well known that measurements with Pt wire probes are less accurate than with Haber-Luggin electrodes as applied in [2]. In the summary of the measured results, presented in Figure 7, this voltage shift has been considered and compensated.

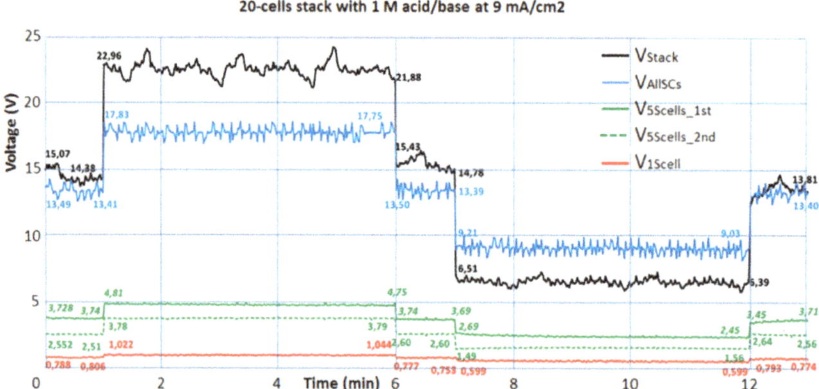

Figure 6. Charge and discharge behavior of a 20-cell stack with 1 M acid/base at 25 °C and 9 mA/cm^2 current density. The black line is the stack voltage measured across the electrodes; the blue line is the stack voltage across the 20-cell units. The voltages measured across the first repeating cell unit next to the electrode (red), the first 5-cell unit (green, solid) and the second 5-cell unit (green, dotted) are also shown.

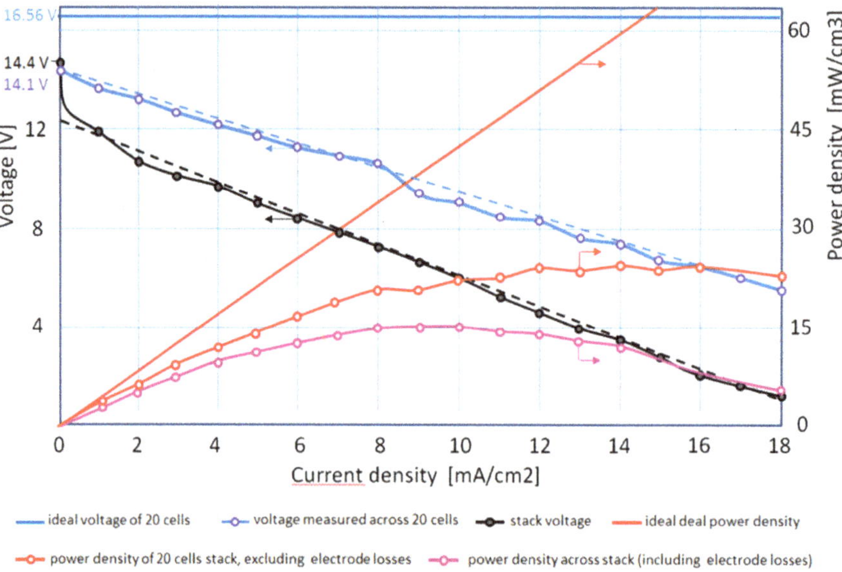

Figure 7. Discharge voltages and net discharge power densities of the 20-cell stack with 1 M acid/base at 25 °C, plotted over discharge current density.

In Figure 8, measured voltages are presented for increasing current densities under charge (increasing voltages) and under discharge (decreasing voltages). At each current density, the measurement continued until a stable voltage was established. Figure 8a shows the results of a 20-cell stack and Figure 8b of a 15-cell stack, for 1M acid and base. Both results show the same general trends. The voltages measured across the electrodes of the whole stack are given by black dots and the measurements across all repeating cell units (excluding the electrodes) by blue dots. In addition, measurements across the first 5-cell unit (next to one electrode) are given by green solid lines and the measurements across the second 5-cell unit (in the middle of the stack) by green broken lines. Measurements for one repeating cell unit next to one electrode are displayed in red. If a shift of +0.7 V is imposed on the blue curves, as discussed before, the electrode losses between charge and discharge are about equal. They increase moderately with increasing current density.

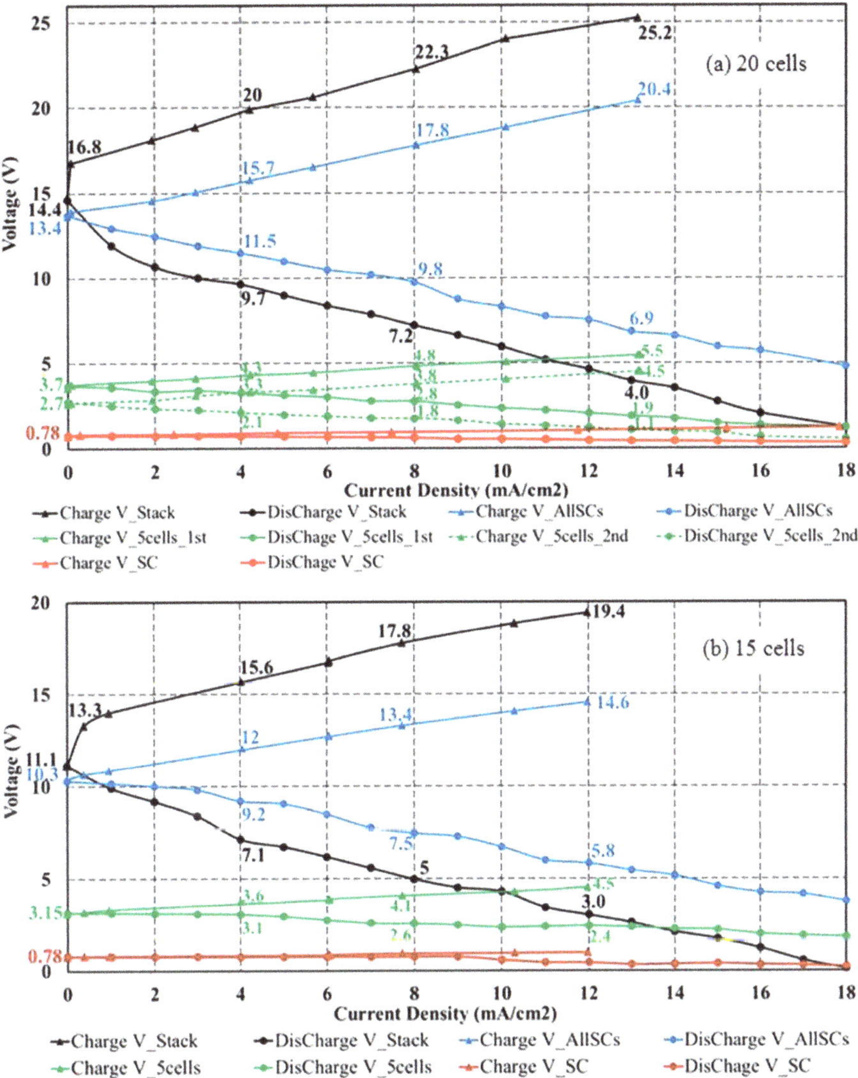

Figure 8. Current density measurements during charge and discharge with 1 M acid and base at 25 °C and 0.1 MPa at steady state. Current-voltage curves of positive slope represent charging, of negative slope discharging. Results for a 20-cell stack are shown in (**a**) and for a 15-cell stack in (**b**). Displayed are the measured voltages, VSC (across BP (bipolar membranes) of the first single cell, red), V5cell_2nd (across the second 5-cell unit, green dotted), V5cell_1st (across the first 5-cell unit, green solid), VAllSCs (across all cell units measured between Pt electrodes, blue) and VStack (stack voltage between electrodes, black).

A comparison of Figure 8a,b shows that the voltage losses increase with the number of unit cells. This is a consequence of the fact that the shunt current accumulates in the stack center, as discussed qualitatively with Figure 5. This shunt current has to be compensated by an increased neutralization reaction of H^+ and OH^- in the bipolar membrane. The unit cells in the stack center therefore contribute less to the stack voltage if the cell number increases. A decrease of the voltage of the second 5-cell unit, located in the middle of the stack, compared to the first 5-cell voltage is also obvious in Figure 8a, demonstrating the influence of the increasing shunt current in the middle of the stack, both at charge and discharge.

3.4. Discharge Power Density and Efficiency

The focus of the present contribution is on the discharge behavior of flow batteries. Here, as in [2], power density PD, defined as net electric power, divided by stack volume, is useful for characterizing the discharge. For the stack volume, only the active area of the membranes and the thickness of each repeating cell unit (1.96 mm) have been considered, while the volume of both electrode chambers has not been included. In Figure 7, discharge voltage and net discharge power density of the tested 20-cell stack with 1 M acid/base at 25 °C are displayed over discharge current density. As explained in Section 3.3, the voltages measured across all single cells (the blue dotted line) have been moved up by 0.7 V (compared to Figure 8), to compensate for measurement errors of the Pt probes.

Considering an ideal behavior without losses across the stack (zero resistances) and with the theoretical single cell voltage of 0.828 V, a constant voltage of 16.56 V would result, leading to a linear increase of power density over current density, marked by the solid red line. The actually measured voltages across all 20 cells of the stack over current density lead to the power densities as displayed by the red line with measurement bullets. The losses partly result from the Ohmic resistances across the membranes and the electrolyte compartments, as already considered in the single-cell experiments in [2]. Another part results from the shunt currents, discussed in Section 3.2.

If we include the measured voltage drop over the electrodes, the pink line with bullets results in the measured power density of the stack. The difference between the red and the pink line is attributed to additional losses in the electrode compartments. The fact that the total losses are substantially higher than predicted from the single-cell experiments can be attributed to the shunt currents through the feed/exit lines. In addition, a nonuniform flow distribution of the electrolytes through adjacent cells could have contributed. This can be concluded from differences between measured voltages of adjacent cell units.

The highest power density for the 20-cell stack of 15 mW/cm^3 has been reached at 10 mA/cm^2, corresponding to a stack power output of 6 W with 36% voltage efficiency. Neglecting the electrode chamber losses, the highest stack power goes up to 9 W or 24 mW/cm^{-3} with 50% voltage efficiency.

The power density decreases at higher current densities due to increasing resistive losses. As observed in the single-cell experiments [2], discharge current densities well above 20 mA/cm^2 resulted in unstable behavior, with voltage breakdown due to water accumulation inside BP. However, this is above the discharge power densities reached in the stack experiments.

4. Conclusions

The experimental results obtained in this study show that the single cell results presented in [2] can be extrapolated to multicell stacks, but additional losses have to be considered. Most importantly, the influence of shunt currents has to be taken into account. As demonstrated with open current experiments under self-discharging conditions in Section 3.2, the influence of shunt currents through the parallel acid/base feed/exit lines increases with increasing number of cell units. It is most pronounced in the middle of the stack, where it can substantially reduce the open cell voltage.

The single cell experiments reported in [2] showed that at higher acid/base concentrations, a breakdown of the discharge voltage occurred, if the current density was raised above a certain value between 20 and 40 mA/cm^2. This was attributed to a delamination of the bipolar membrane, caused by the excessive formation of water through the neutralization reaction of H^+ and OH^- in the reaction layer. However, these current densities were well above the discharge current densities reached in the stack experiments of Figures 6 and 8. Delamination of bipolar membranes during discharge was not observed in the stack results presented.

Figure 7 summarizes the experimental discharge results for the 20-cell stacks and points to possibilities for improvement. A reduction of the influence of shunt currents would directly reduce the difference between the (ideal) red solid line and the measured red dotted line. A straightforward approach would be to change the acid and base flow pattern from parallel to a serial flow, where all of the acid and base flows consecutively upwards in one and downwards in the next acid/base chamber. Then, minor shunt currents could only occur between adjacent cells, driven by the single cell voltages but not by the total stack voltage, as in case of the parallel feed/exit lines. Preliminary experiments at open current under self-discharge (similar to Figure 4) proved that the self-discharge of a 5-cell stack with serial flow of acid and base was considerably slower, compared with a similar stack with parallel flow. However, for the present design the pressure drop across the acid/base feed/exit lines was substantially increased. This would result in an unacceptable pressure drop for a 20-cell stack with serial flow.

Since Figure 4 has shown that a 5-cell stack with parallel flow was only marginally affected by shunt currents, a combination of parallel and serial flow could be a compromise between pressure drop and shunt current. The parallel flow through the acid and base compartments should change direction after, for example, every 5 to 10 cells. Electrode losses (the difference between the red and the pink lines in Figure 7) could be reduced, e.g., by improved, gas-consuming electrodes [13].

With respect to higher power densities, the positive influence of larger concentrations of the electrolytes has to be balanced against reduced permselectivities of the ion exchange membranes used. Here, the challenge is to develop monopolar and bipolar membranes with increased fixed-charge concentrations as already discussed in [2]. In the single-cell experiments [2], water accumulation at the BP interface resulted in a breakdown of the cell voltage at current densities above about 30 mA/cm^2 for 1M acid/base. This was well above the current densities reached in the stack experiments. Nevertheless, an improved water transport through BP will be a further challenge for BP membranes optimized for REDBP.

Author Contributions: Conceptualization, G.E., H.S. and U.N.; formal analysis, J.X.; investigation, J.X.; methodology, J.X. and H.S.; resources, J.X. and H.S.; supervision, G.E., H.S. and U.N. All authors have read and agreed to the published version of the manuscript.

Funding: Jiabing Xia gratefully acknowledges the Ph.D. scholarship from the GREES program at University of Stuttgart in Germany.

Conflicts of Interest: The authors declare no conflict of interest.

List of Abbreviations

AM	anion exchange membrane
BM	bipolar membrane
CM	cation exchange membrane
OCV	open circuit voltage
REDBP	reverse electrodialysis with bipolar membranes

References

1. Post, J.W.; Veerman, J.; Hamelers, H.V.M.; Euverink, G.J.W.; Metz, S.J.; Nymeijer, K.; Buisman, C.J.N. Salinity-gradient power: Evaluation of pressure retarded osmosis and reverse electrodialysis. *J. Membr. Sci.* **2007**, *288*, 218–230. [CrossRef]
2. Xia, J.; Eigenberger, G.; Strathmann, H.; Nieken, U. Flow battery based on reverse electrodialysis with bipolar membranes: Single cell experiments. *J. Membr. Sci.* **2018**, *565*, 157–168. [CrossRef]

3. Walther, J.F. Process for Production of Electrical Energy from the Neutralization of Acid and Base in a Bipolar Membrane Cell. U.S. Patent 06183483, 19 January 1982.
4. Zholkovskij, E.; Müller, M.; Staude, E. The storage battery with bipolar membranes. *J. Membr. Sci.* **1998**, *141*, 231–243. [CrossRef]
5. Pretz, J.; Staude, E. Reverse electrodialysis (RED) with bipolar membranes, an energy storage system. *Phys. Chem.* **1998**, *102*, 676–685. [CrossRef]
6. Huang, C.; Xu, T. Electrodialysis with bipolar membranes for sustainable development. *Environ. Sci. Technol.* **2006**, *40*, 5233–5243. [CrossRef] [PubMed]
7. Parasuraman, A.; Lim, T.; Menictas, C.; Skyllas-Kazacos, M. Review of material research and development for vanadium redox flow battery applications. *Electrochim. Acta* **2013**, *101*, 27–40. [CrossRef]
8. Xia, J. Reverse Electrodialysis with Bipolar Membranes (REBP) as an Energy Storage System. Ph.D. Thesis, Fakultät Verfahrenstechnik of Stuttgart University, Stuttgart, Germany, 10 August 2018.
9. Fumatech, Home Page. Available online: www.fumatech.com (accessed on 10 January 2020).
10. Yin, C.; Guo, S.; Fang, H.; Liu, J.; Li, Y.; Tang, H. Numerical and experimental studies of stack shunt current for vanadium redox flow battery. *Appl. Energy* **2015**, *151*, 237–248. [CrossRef]
11. Fink, H.; Remy, M. Shunt currents in vanadium flow batteries: Measurement, modelling and implications for efficiency. *J. Power Sources* **2015**, *284*, 547–553. [CrossRef]
12. Ye, Q.; Hu, J.; Cheng, P.; Ma, Z. Design trade-offs among shunt current, pumping loss and compactness in the piping system of a multi-stack vanadium flow battery. *J. Power Sources* **2015**, *296*, 352–364. [CrossRef]
13. Kintrup, J.; Millaruelo, M.; Trieu, V.; Bulan, A.; Mojica, E.S. Gas Diffusion Electrodes for Efficient Manufacturing of Chlorine and Other Chemicals. *Electrochem. Soc.* **2018**, *26*, 73–76. [CrossRef]

Article

Facile Synthesis of Bio-Template Tubular MCo_2O_4 (M = Cr, Mn, Ni) Microstructure and Its Electrochemical Performance in Aqueous Electrolyte

Deepa Guragain [1], Camila Zequine [2], Ram K Gupta [2] and Sanjay R Mishra [1,*]

1 Department of Physics and Materials Science, The University of Memphis, Memphis, TN 38152, USA; ddeepag13@gmail.com
2 Department of Chemistry, Pittsburg State University, Pittsburg, KS 66762, USA; camilazequine@hotmail.com (C.Z.); rgupta@pittstate.edu (R.K.G.)
* Correspondence: srmishra@memphis.edu

Received: 17 January 2020; Accepted: 10 March 2020; Published: 16 March 2020

Abstract: In this project, we present a comparative study of the electrochemical performance for tubular MCo_2O_4 (M = Cr, Mn, Ni) microstructures prepared using cotton fiber as a bio-template. Crystal structure, surface properties, morphology, and electrochemical properties of MCo_2O_4 are characterized using X-ray diffraction (XRD), gas adsorption, scanning electron microscopy (SEM), Fourier transforms infrared spectroscopy (FTIR), cyclic voltammetry (CV), and galvanostatic charge-discharge cycling (GCD). The electrochemical performance of the electrode made up of tubular MCo_2O_4 structures was evaluated in aqueous 3M KOH electrolytes. The as-obtained templated MCo_2O_4 microstructures inherit the tubular morphology. The large-surface-area of tubular microstructures leads to a noticeable pseudocapacitive property with the excellent electrochemical performance of $NiCo_2O_4$ with specific capacitance value exceeding 407.2 F/g at 2 mV/s scan rate. In addition, a Coulombic efficiency ~100%, and excellent cycling stability with 100% capacitance retention for MCo_2O_4 was noted even after 5000 cycles. These tubular MCo_2O_4 microstructure display peak power density is exceeding 7000 W/Kg. The superior performance of the tubular MCo_2O_4 microstructure electrode is attributed to their high surface area, adequate pore volume distribution, and active carbon matrix, which allows effective redox reaction and diffusion of hydrated ions.

Keywords: bio-template; MCo_2O_4 (M = Cr, Mn, Ni); electrochemical; cyclic voltammetry; specific capacitance

---

1. Introduction

Supercapacitors (SCs) are the energy storage device. SCs are in high demand because of their greater power density compared to batteries and higher energy density compared to that of capacitors [1]. In the capacitor, there is no time lag during the charging process; hence it can give higher power density, and in the battery, there is low self-discharge so that it can provide higher energy density [2]. Because of this, a supercapacitor is easy to charge within a short time and able to show significant performance even after prolonged use. SCs are of three types: (i) electric double-layer capacitors (EDLCs), (ii) pseudo capacitors (PCs), and (iii) hybrid capacitors. EDLCs are based on the principle that physical adsorption takes place on the interface of a solid electrode, usually carbon-based material, and liquid electrolytes [3,4]. In PCs, surface redox reaction takes place at the electrode-electrolyte interface, which is responsible for storing electronic charges [5], where metal oxides and conducting polymer-based materials are used as active electrode materials [6,7]. EDLCs have lower specific capacitance and energy density as compared to PCs, hence, practically PCs are in higher demand. A hybrid capacitor is a combination of both EDLCs and PCs; examples are carbon nanotubes, graphene, etc. [8,9]. They display hybrid charge–storage mechanisms and have the ability to deliver higher capacity [10].

The electrolyte ion transport in supercapacitor devices occurs through an ion-transport layer separated from the electrode. The charge storage mechanism follows at the electrode surface during the charging-discharging process [11].

Transition metal oxides (TMOs) with novel nano-architectures and rich in redox reactions are increasingly explored for their application in energy storage devices [12]. Among these, cobalt oxides are highly attractive because of their higher theoretical value for specific capacitance, i.e., 3560 F/g [13]. The nanoarchitecture of these metal-oxides is controlled by the synthesis route, which often requires complex technological strategies, including toxic organic reagents, which might make it difficult for their mass industrial application. Thus, it is highly sought to explore cost-effective facile synthesis strategies and environmentally benign techniques for preparing these electrodes. Furthermore, the ideal electrode should have a high specific surface area for better specific capacitance, controlled porosity for better rate capability as well as specific capacitance, and higher electronic conductivity to improve rate capability and power density of supercapacitor. Nowadays, the bio-templating technique has emerged as a convincing technique for the preparation of oxide supercapacitors [14–17]. Nature offers rich and diverse bio-templates [18–20] like bamboo, lotus pollen grains [21], leaf [22], sorghum straw [23], butterfly wing [24], jute fibers [25], and cotton [12]. Such bio-templates offer elaborate interior and exterior surfaces, and geometries, which make these templates attractive materials to produce multiscale hybrid and hierarchical morphologies.

Numbers of research are published on the study of transition metal oxide-based electrodes such as $MnCo_2O_4$ [5,26–28] and $NiCo_2O_4$ [29–33], Co_3O_4 [34], and $NiFe_2O_4$ [35] for their supercapacitive application. The benefit of transition metal oxides as electrode materials are innumerable, as their multiple oxidation states facilitate multiple redox reactions during electrochemical reaction vis-a-vise offers stable structure. The co-existence of two different cations provides abundant active sites to perform fast reversible faradic redox reaction on the electrode interface; as a result, higher specific capacity, and excellent rate capability are achieved [25,36]. Additionally, the types of bonds between transition metal ions and ligands are dictated by electronegativity and ionization energies [37]; with the former, the structure is dense, while with later the structure is more open. The valence state, ionic radius, electronegativity [38], and the local environment of the cations are affected by the change in Gibbs free energy and electrochemical potential of the electrode. An increase in the electrochemical potential of cathodes is observed with the increase in the number of electrons in *d* orbitals of transition metal elements. This implies a higher consumption of energy during electron transfer [39]. In mixed transition metal oxides, there is a synergetic effect between metal cations; this produces higher electrical conductivity of single metal oxide where there is low activation energy to transfer electron between metal cations and gives excellent structural stability [40,41].

Here we present a comparative study to understand the electrochemical behavior of MCo_2O_4 (M = Cr, Mn, and Ni) electrodes prepared via a facile bio-template method. The electronegativity differences among M ions viz. Cr (1.66), Mn (1.55), Ni (1.99), and Co (1.88) could have a substantial effect on the electrochemical performance of the said electrodes. The electronegativity difference determines the structure, covalent vs. ionic, and the electric potential of the electrode for the charge transfer, as discussed above. In the present study, MCo_2O_4 electrode material is prepared via the bio-template method, where the product assumes the morphology of the microstructure of bio-template and ends up with a carbon matrix. The bio-template method adapted to produce active material inherently fixed in the carbon matrix. The carbon matrix is known to enhance electrode electrochemical performance [42]. The template supported mineralization MCo_2O_4 at room temperature (RT) produces 3D-hierarchical and porous-MCo_2O_4 superstructures with tubular-like morphologies. The doped Co_3O_4 (MCo_2O_4) is explicitly explored in this study as dopant atoms or vacancies are known to affect the crystal field [43], thus modifying the electronic structure and adjusting the electrochemical potential [44].

2. Experimental

2.1. Synthesis

All the chemicals required for the synthsis, such as Cobalt nitrate hexahydrate; $Co(NO_3)_2.6H_2O$, Chromium nitrate hexahydrate; $Cr(NO_3)_2.6H_2O$, Manganese nitrate hexahydrate; $Mn(NO_3)_2.6H_2O$, and Nickel nitrate hexahydrate; $Ni(NO_3)_2.6H_2O$ were purchased from Sigma-Aldrich, St. Louis, Missouri, USA. The spinel MCo_2O_4 (M = Cr, Mn, Ni) was synthesized by a facile bio-template method. Quantities of 1.16 g of $Co(NO_3)_2.6H_2O$ and 0.8 gm of $Cr(NO_3)_2.6H_2O$, 0.55 g of $Co(NO_3)_2.6H_2O$ and 0.17 gm of $Mn(NO_3)_2.6H_2O$, and 0.86 g of $Co(NO_3)_2.6H_2O$ and 0.43 gm of $Ni(NO_3)_2.6H_2O$ were mixed in 15 ml of distilled water separately, and the mixture was ultra-sonicated for 10 minutes to make a homogenous solution. Then, 1.0 g of cotton was soaked in the mixture solutions for 5 minutes. The resulting soaked cotton was filtered and dried at 150 °C for 30 minutes. The dried cotton was later calcined at 520 °C for 3 hours in the air to obtain bio-templated $CrCo_2O_4$, $MnCo_2O_4$, and NiCo2O4 tubular microstructure.

2.2. Characterization

The x-ray diffraction patterns were obtained via Bruker D8 Advance X-ray diffractometer (Bruker Corporation, Madison, WI, USA) using Cu K_α radiation to check phase purity and determine the crystalline parameters of as-prepared samples. A scanning electron microscope (Phenom) at 10 keV analyzed the morphology of samples. The Brunauer–Emmett–Teller (BET) method was used to measure the specific surface area of the samples. The surface area measurement was carried out by adsorption-desorption isotherms at 77 K, (Quantachrome, Boynton Beach, FL 33426, model No. AS1MP) using nitrogen as adsorbing gas. Thermogravimetric analyses (TGA, Instrument Specialist, Inc., Twin Lakes, WI, USA), were performed in 24 to 550 °C temperature range. FTIR spectra were collected via Theromo-Fisher Scientific FTIR spectrometer (Nicolet iS10, Thermo Fisher Scientific, Waltham, MA, USA) between 450 and 1000 cm^{-1}.

Versastat 4–500 electrochemical workstation (Princeton Applied Research, USA) was used to perform electrochemical measurements in a standard three-electrode configuration. To prepare an electrode, slurry pastes of 80 wt % of the synthesized powder, 10 wt % of acetylene black, and 10 wt % of polyvinylidene difluoride (PVDF) were mixed in the presence of N-methyl pyrrolidinone (NMP). The thoroughly mixed paste was applied onto a nickel foam. Here, the active mass is 80% out of the total pasted mass in the electrode. The prepared electrodes were dried under vacuum at 60 °C for 10 hours. The loading mass of all samples was about 2–3 mg, measured by weighing the nickel foam before and after deposition with an analytical balance (MS105DU, Mettler Toledo, 0.01 mg of resolution). MCo_2O_4 (M = Cr, Mn, Ni) coated nickel foam was used as a working electrode, a saturated calomel electrode (SCE) as a reference electrode, and a platinum wire as a counter electrode. The electrochemical performance of the electrodes was evaluated at RT in 3M KOH electrolyte via cyclic voltammetry and galvanostatic charge-discharge techniques measurements.

3. Results and Discussion

Figure 1a shows the XRD patterns of the bio-templated $CrCo_2O_4$, $MnCo_2O_4$, and $NiCo_2O_4$ microstructure. The XRD patterns match with the face-centered cubic phase of $CrCo_2O_4$, $MnCo_2O_4$, and $NiCo_2O_4$ (International Centre for Diffraction Data (ICDD) #02-0770). The main peaks at 30.9°, 36.4°, 44.3°, 58.6°, and 64.3° for $CrCo_2O_4$, 31.1°, 36.7°, 44.7°, 59.2°, and 65.9° for $MnCo_2O_4$, and 31.3°, 36.8°, 44.8°, 59.4°, and 65.2° for $NiCo_2O_4$ can be assigned to the (220), (311), (400), (511) and (440) reflections of $CrCo_2O_4$, $MnCo_2O_4$, and $NiCo_2O_4$ respectively [45,46]. The pattern of $NiCo_2O_4$ shows a peak at 43.2°, which indicates the formation of NiO cubic phase as also confirmed by TOPAS fitting. The lattice constants obtained using d-spacing for the sample are $a = b = c =$ 0.816 nm, 0.808 nm, 0.807 nm, and for $CrCo_2O_4$, $MnCo_2O_4$, and $NiCo_2O_4$, respectively. The crystallite size of $CrCo_2O_4$, $MnCo_2O_4$, and $NiCo_2O_4$ as calculated using Scherrer's formula [47] is around 10.57 nm, 14.65 nm, and 19.97 nm for

$CrCo_2O_4$, $MnCo_2O_4$, and $NiCo_2O_4$, respectively (Table 1). FTIR spectrum, Figure 1b, further identifies the structure of the bio-templated MCo_2O_4. The FTIR spectrum displays two distinct bands at 515.7 (ν_1) and 637 (ν_2) cm^{-1}, which arise from the stretching vibrations of the metal-oxygen bonds [48–50]. The ν_1 band is characteristic of M-O (M = Cr, Mn, Ni) vibrations in octahedral coordination, and the ν_2 band is attributable to M-O (M - Co) bond vibration in tetrahedral coordination. These frequency bands are the signature vibrational bands for the spinel lattice [51]. Hence FTIR spectrum at 519.1, 519.02, and 519.08 cm^{-1} indicate stretching vibration of Co^{3+}-O^{2-} in the octahedral sites, and at 638.6, 639.9, and 641.3 cm^{-1} indicate vibration of Cr^{3+}-O^{-}, Mn^{2+}-O^{2-}, and Ni^{2+}-O^{-} at tetrahedral sites for $CrCo_2O_4$, $MnCo_2O_4$, and $NiCo_2O_4$, respectively [52]. The presence of vibration bands confirms the development of pure phase spinal $CrCo_2O_4$, $MnCo_2O_4$, and $NiCo_2O_4$ nanostructures.

Table 1. Crystallite size and physical properties of MCo_2O_4 (M = Cr, Mn, Ni) determined using XRD, the Barrette–Joyner–Halenda (BJH) method, and Brunauer–Emmett–Teller (BET) surface area analyzer.

Sample	BET Surface Area(m^2/g)	BJH Surface Area (m^2/g)	BJH Avg. Pore Radius (nm)	BJH Avg. Pore Volume (cc/g)	Crystallite Size (nm)
$CrCo_2O_4$	34.4	46.9	1.135	0.106	10.57
$MnCo_2O_4$	32.2	47.7	0.839	0.071	14.65
$NiCo_2O_4$	18.9	31.8	1.129	0.039	19.97

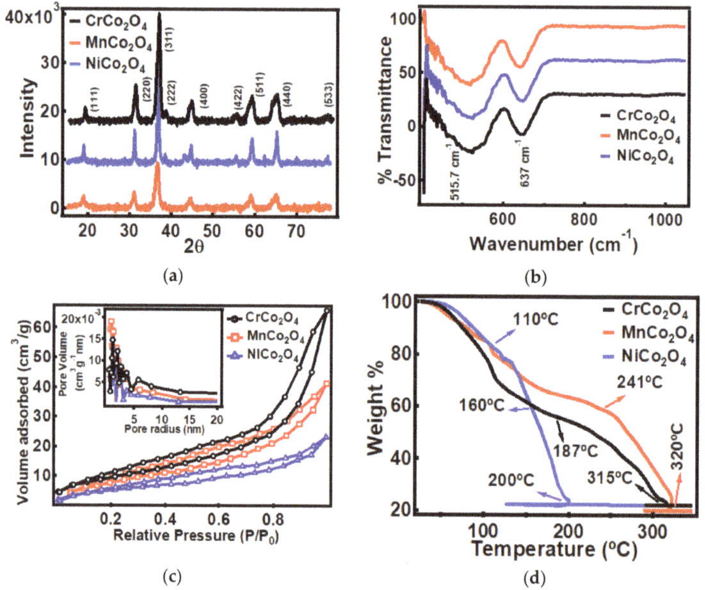

Figure 1. (**a**) x-ray diffraction pattern, (**b**) FTIR, (**c**) adsorption-desorption curve and inset pore volume distribution, and (**d**) thermogravimetric curve of tubular MCo_2O_4 (M = Cr, Mn, Ni) structures.

Figure 1c shows the BET specific surface area of tubular MCo_2O_4 microstructures. The specific surface area was determined from N_2 adsorption-desorption isotherms obtained at 77 K between relative pressure P/P_o~0.029 to 0.99, and the Barrette–Joyner–Halenda (BJH) method was used for measuring corresponding pore sized distributions. The type IV isotherm hysteresis loops [53] suggest the existence of mesopores in the samples. The BET specific surface area of biomorphic $CrCo_2O_4$, $MnCo_2O_4$, and $NiCo_2O_4$ are 34.4 m^2/g, 32.2 m^2/g, and 18.9 m^2/g, respectively. Figure 1c inset shows the pore size distribution. Inset curves indicate having a more favorable condition for the fast ion transport phenomenon within the electrode surface [54–57], which is confirmed by the presence of a significant number of pores distribution at around 0.4 nm to 4.3 nm with the highest pore volume.

Additionally, the large BET surface area of tubular MCo_2O_4 superstructures can provide plenty of superficial electrochemical active sites to participate in the Faradaic redox reactions.

The thermogravimetric analysis was conducted on the infiltrated samples (cotton dipped in a mixture of chemical solution and filtered it) to understand the temperature dependence mechanism of the formation of biomorphic MCo_2O_4. Figure 1d shows the TGA plots of MCo_2O_4 measured in the temperature range of 24 to 550 °C. The formation of MCo_2O_4 from the nitrate salts results in three steps. The weight loss at around 110 °C for all three MCo_2O_4 is due to water desorption, the second weight loss up to 187 °C for $CrCo_2O_4$, 241 °C for $MnCo_2O_4$, and 160 °C for $NiCo_2O_4$ is due to burning of cotton and start of decomposition of $Co(NO_3)_2{\cdot}6H_2O$ and $Cr(NO_3)_2{\cdot}6H_2O$, $Co(NO_3)_2{\cdot}6H_2O$ and $Mn(NO_3)_2{\cdot}6H_2O$, $Co(NO_3)_2{\cdot}6H_2O$ and $Ni(NO_3)_2{\cdot}6H_2O$ respectively, there was no weight loss at beyond 315, 320, and 200 °C which signifies the completion of the formation of $CrCo_2O_4$, $MnCo_2O_4$, and $NiCo_2O_4$. Upon immersing fiber into the precursor solution, the water and $Cr(NO_3)_2{\cdot}6H_2O$, $Mn(NO_3)_2{\cdot}6H_2O$, $Co(NO_3)_2{\cdot}6H_2O$ and $Ni(NO_3)_2{\cdot}6H_2O$ molecules were absorbed onto the hydroxyl-group-rich cotton fiber substrate. With the heat treatment above 520 °C, nitrate salts decomposed in the form $CrCo_2O_4$, $MnCo_2O_4$, and $NiCo_2O_4$ as follow [58],

$$2Co(NO_3)_2{\cdot}6H_2O + Cr(NO_3)_2{\cdot}6H_2O \rightarrow CrCo_2O_4 + 2O_2 + 6NO_2 + 18H_2O \quad (1)$$

$$2Co(NO_3)_2{\cdot}6H_2O + Mn(NO_3)_2{\cdot}6H_2O \rightarrow MnCo_2O_4 + 2O_2 + 6NO_2 + 18H_2O \quad (2)$$

$$2Co(NO_3)_2{\cdot}6H_2O + Ni(NO_3)_2{\cdot}6H_2O \rightarrow NiCo_2O_4 + 2O_2 + 6NO_2 + 18H_2O \quad (3)$$

With the increase in calcination temperature, the removal of organic substance was achieved where the remaining few portions of the organic substance change into carbon.

FE-SEM in Figure 2a–d displays tubular morphology of the cotton fibers, samples $CrCo_2O_4$, $MnCo_2O_4$, and $NiCo_2O_4$, respectively, which resembles a biomorphic structure. Figure 3a–c shows SEM images obtained using elemental mapping at chromium, manganese, and nickel energy peaks and shows that the tubular structure is well decorated with the $CrCo_2O_4$, $MnCo_2O_4$, and $NiCo_2O_4$ nanoparticles. Table 2 gives the elemental composition and element distribution, which is obtained via EDX (energy dispersive x-ray spectroscopy).

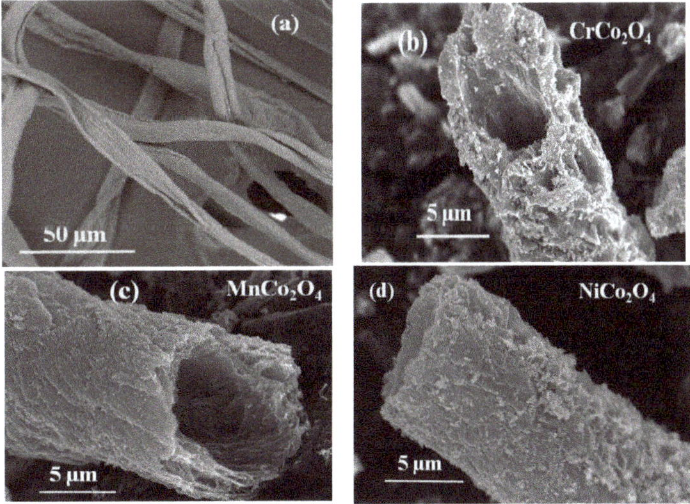

Figure 2. SEM images of (**a**) cotton fiber, (**b**), (**c**), and (**d**) bio-templated tubular MCo_2O_4 (M = Cr, Mn, Ni) structures.

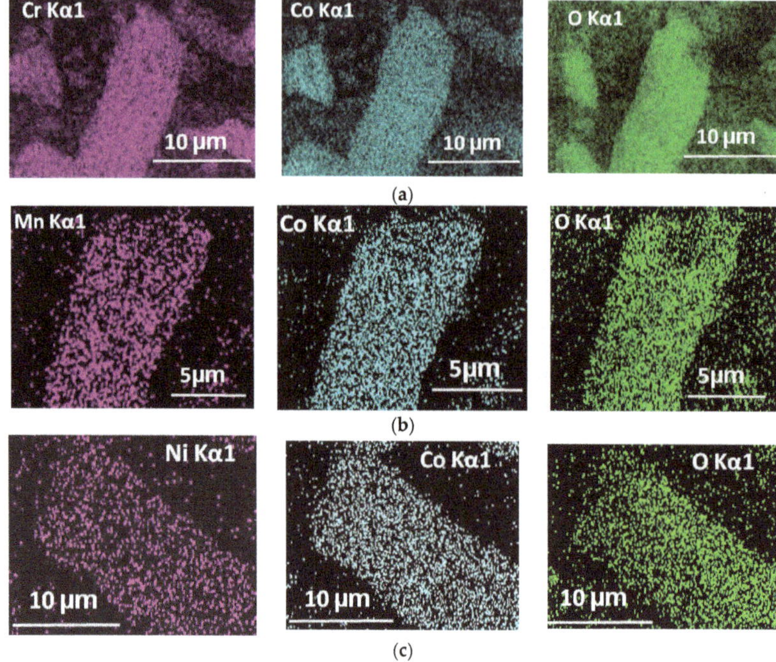

Figure 3. (**a**), (**b**), and (**c**) shows EDX mapping of tubular MCo_2O_4 (M = Cr, Mn, Ni) structures, respectively.

Table 2. Elemental composition in wt % for MCo_2O_4 (M = Cr, Mn, Ni) obtained using energy dispersive X-Ray analysis (EDX). The elemental composition is approximately determined using EDX. Ideally, EDX can prove which elements are abundant in the particles, but not obtain the exact chemical composition.

	Co	C	O	Cr	Mn	Ni
$CrCo_2O_4$	19.5	49.4	23.2	7.8		
$MnCo_2O_4$	33.1	34.6	23.5		8.8	
$NiCo_2O_4$	27.2	42.4	19.3			11.1

The type of electrolytes and their molar concentration play a vital role in determining the electrochemical behavior of oxide electrodes [59–61]. Therefore, many aqueous electrolytes such as sulfates K_2SO_4, H_2SO_4, KNO_3, Na_2SO_4, hydroxyl KOH, NaOH, LiOH, and chlorides KCl, NaCl have been explored to be used in supercapacitors [62–66]. The ultimate performance of the electrode is based on the properties of the electrode material and the intercalation efficiency of the cations [51]. Since KOH electrolyte provides lower electrochemical series resistance with better conductivity as compared to other electrolytes [67], KOH is chosen as an electrolyte in this study for the electrochemical measurement.

Cyclic voltammetry and charge-discharge curves were measured to investigate the electrochemical behavior of MCo_2O_4 nanoparticles. Figure 4 displays the CV curves for tubular MCo_2O_4 electrodes measured in the 3M KOH electrolyte. Figure 4a,c, and Figure 4e shows the CV curves measured in the voltage window of 0.0 to 0.6 V and measured at different scan rate from 2 to 300 mV/s. A pair of redox peaks associated with the redox reactions involved in the alkaline electrolyte during the charging and discharging process was observed in all CV plots. The CV curve is asymmetric, which indicates a quasi-reversible redox reaction [68], the anodic and cathodic peak separation are 0.121 V, 0.124 V, and 0.123 V at 2 mV/s and 0.310 V, 0.186 V, and 0.333 V at 300 mV/s for $CrCo_2O_4$, $MnCo_2O_4$ and $NiCo_2O_4$ respectively. The presence of anodic and cathodic peaks, indicating the usefulness of the materials as a pseudocapacitor. Typical pseudo-capacitance behavior of MCo_2O_4 nanostructures arises from the reversible surface or near-surface Faradic reactions for charge storage. The reversible redox reaction involved in the charge-discharge process for MCo_2O_4 can be described as follows by Equations (5)–(7) [69–71].

$$MO + OH^- \leftrightarrow MOOH + e^- \tag{4}$$

$$CrCo_2O_4 + OH^- + H_2O \leftrightarrow CrOOH + 2CoOOH + e^- \tag{5}$$

$$MnCo_2O_4 + OH^- + H_2O \leftrightarrow MnOOH + 2CoOOH + e^- \tag{6}$$

$$NiCo_2O_4 + OH^- + H_2O \leftrightarrow NiOOH + 2CoOOH + e^- \tag{7}$$

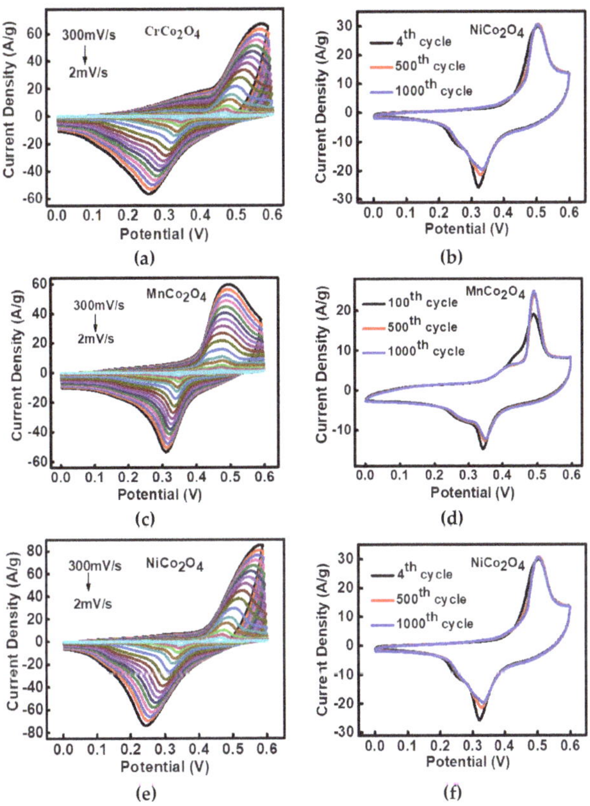

Figure 4. (**a**,**c**,**e**) show cyclic voltammetry curves of tubular MCo_2O_4 (M = Cr, Mn, Ni) electrode obtained in the scan range of 5 mV/s to 300 mV/s measured in 3M KOH electrolyte. (**b**,**d**,**f**) show cyclic stability curves measured up to 1000 cycles in 3M KOH electrolyte at scan rate of 40 mV/s.

Pseudocapacitive characteristics of electrodes are indicated by a non-rectangular form of CV curves. Within the potential range from 0 to 0.6 V, a pair of reversible redox peaks can be observed. With the increase in the scan rate, a small positive shift of the oxidation peak potential and a negative shift of the reduction peak potential was observed, which can be primarily attributed to the influence of the increasing electrochemical polarization as the scan rate scales up. Pairs of reversible redox curve are indicative of pseudocapacitive behavior of the material with redox peaks attributed to M(II)/M(III) redox process [72]. The redox potentials and shape of the CV curves are comparable to those reported for $CrCo_2O_4$, $MnCo_2O_4$, and $NiCo_2O_4$ electrodes [24,73–76], suggesting that the measured capacitance mainly arises from the redox mechanism.

Figure 5a shows the specific capacitance, C_{sp}, as a function of the voltage scan rate of the tubular MCo_2O_4 electrode. The specific capacitance, C_{sp}, was calculated from the CV plots using the following Equation (8) [77].

$$C_{sp} = \frac{\int_{V1}^{V2} i * V * dV}{m * v * (V2 - V1)} \tag{8}$$

where V_1 and V_2 stand for the working potential limits, i stands for the current, m stands for the mass of the electroactive materials, and v is the scan rate in mV/s.

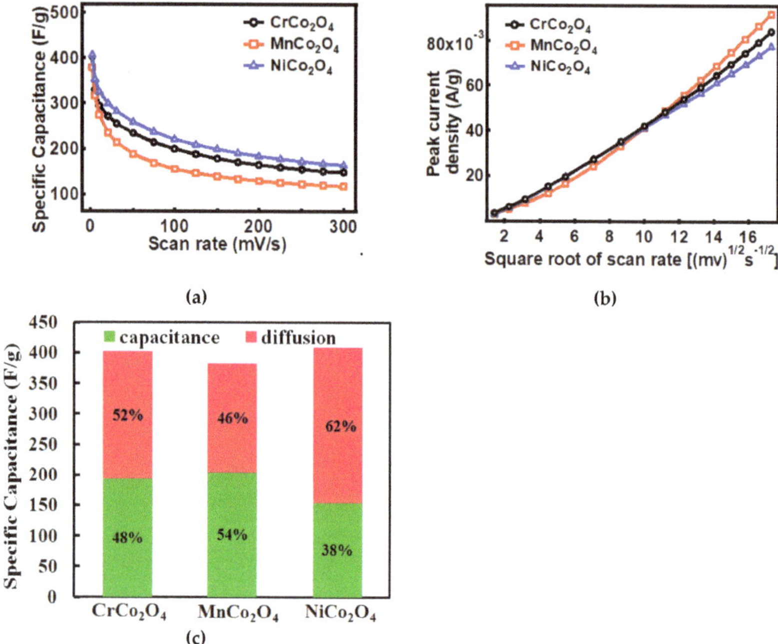

Figure 5. (**a**) Specific capacitance vs. scan rate, (**b**) and peak current vs. (scan rate)$^{1/2}$, and (**c**) diffusion and capacitive contribution to the specific capacitance.

It is evident for this figure that the electrode displays higher C_{sp} up to 407.2 F/g for $NiCo_2O_4$ in 3M KOH electrolyte at 2 mV/s, which is higher than the value that is observed for either $CrCo_2O_4$ (C_{sp} ~ 403.2 F/g) or $MnCo_2O_4$ (C_{sp} ~ 378.1 F/g) electrode, value are given in Table 3. The specific capacitance for higher scan rates (>50 mV/s) remains practically constant because of limited ion movement only at the surface of the electrode material. Hence EDLC becomes a dominant mechanism at higher scan rates. At lower scan rates (<5 mV/s), the majority of active surface are utilized by the ions for charge storage, and hence resulting in the higher specific capacitance. The CV cyclic stability of the electrode was tested for 1000 cycles. Figure 4b,d, and Figure 4f show no significant differences in the CV curves after the 100th, 500th, and 1000th cycle of repetition. The CV curves clearly show that the current response is proportionally increased with the scan rate, indicating an excellent capacitive behavior of the electrode materials. This can be ascribed to facile ion diffusion and large specific surface area of the electrode materials. Furthermore, there is almost no relation between the shape of CV curves and scan rates, which can be associated with the electron conduction and improved mass transportation of electrode material [78].

Table 3. Data of specific capacitance, energy density, and power density for MCo_2O_4 (M = Cr, Mn, Ni) obtained from cyclic voltammetry and charge-discharge curves.

	$CrCo_2O_4$	$MnCo_2O_4$	$NiCo_2O_4$
Specific Capacitance at 2mv/s	403.2 F/g	378.1 F/g	407.2 F/g
Specific Capacitance at 1A/g	231 F/g	161 F/g	190 F/g
Energy density	11.1 Wh/Kg	7.8 Wh/Kg	9.3 Wh/Kg
Power density	7287.34 W/Kg	7195.33 W/Kg	7186.12 W/Kg

The total stored charge has a contribution from three components; first is the Faradaic contribution coming from the insertion process of electrolyte ions, second is the faradaic contribution from the charge-transfer process with surface atoms, and third is pseudocapacitance and nonfaradic contribution from the double layer effect [79]. Both pseudocapacitance and double-layer charging are substantial, due to their higher surface area of nanoparticles. The capacitive effects are characterized by analyzing the cyclic voltammetry data at various scan rates according to [80,81],

$$i = av^b \tag{9}$$

where i, v, a and b, are peak current (A), voltage scan rate (mV/s), and fitting parameters, respectively. The charge storage mechanism is defined based on the value of the constant b, where $b = 1$ defines capacitive or $b = 0.5$ defines diffusion-limited charge storage mechanism. Fitting the peak current, i, vs. square root of the scan rate, SQRT (scan rate), $v^{-1/2}$, curves, Figure 5b, with Equation (6), gives b values of ~ 0.646, 0.711, and 0.648 for $CrCo_2O_4$, $MnCo_2O_4$, and $NiCo_2O_4$, respectively. This obtained b value for our sample MCo_2O_4 (M = Cr, Mn, Ni) indicates the diffusive nature of the charge storage mechanism is prominent for $NiCo_2O_4$ as compared to the other two.

Usually, the contribution to the current response at fixed potential comes from surface capacitive effects and diffusion-controlled insertion processes [82,83]. These contributions to the specific capacitance could be separated using the following Equation (10):

$$C_{sp} = k_1 + k_2\, v^{-1/2} \tag{10}$$

For which k_1 and k_2 can be determined from the C_{sp} $vs.$ $v^{-1/2}$ linear plot with slope k_2 and intercept k_1. k_1 and k_2 are fractions of diffusion and capacitive contribution to the net specific capacitance at a given voltage rate. The C_{sp} was plotted against the slow scan rate up to 20 mV/s, and a regression fit was performed using Equation (10). The obtained k_1 and k_2 values were used to determine the fractional contribution to the net specific capacitance. Figure 5c shows capacitive and diffusive fractional contributions to net specific capacitance for a slow scan rate of up to 20 mV/s. By comparing the lower green area with the total capacitance, we find that capacitive effects contribute by 48%, 54%, and 38% of the total specific capacitance for $CrCo_2O_4$, $MnCo_2O_4$, and $NiCo_2O_4$, respectively.

Figure 6a,c, and Figure 6e show the galvanostatic charge-discharge (GCD) plots measured in the voltage window of 0.0 to 0.6 V at different current densities between 0.75 A/g to 30 A/g in 3M KOH. From the observed non-linearity between the potential and time, it is confirmed that the capacitance of the studied materials is not constant over the studied potential ranges. The specific capacitance of electrodes was calculated using the following Equation (11):

$$C_{sp} = \frac{I * t}{m * \Delta V} \tag{11}$$

where C_{sp}, I, ΔV, m, and t are the specific capacitance (F/g), charge-discharge current (A), the potential range (V), and the mass of the electroactive materials, and the discharging time (s), respectively. The GCD curves with a plateau, usually displayed by oxide electrodes, show pseudocapacitive behavior of electrode with respect to their discharging time for all electrolytes. This typical GCD behavior could arise from the electrochemical adsorption-desorption of OH^- electrolyte and/or a redox reaction at the interface of electrode/electrolyte [84,85]. It is observed that the discharging time in biomorphic MCo_2O_4 is longer for $CrCo_2O_4$ in the KOH electrolyte. The specific capacitances of biomorphic $CrCo_2O_4$, $MnCo_2O_4$, and $NiCo_2O_4$ at 1 A/g are 231 F/g, 161 F/g, and 190 F/g in 3M KOH electrolytes, respectively are shown in Table 3. Figure 7a shows the dependence of current density on the specific capacitance of the electrode material. Usually, insufficient Faradic redox reaction is achieved at the high discharge current densities. This leads to increased potential drop due to the resistance of tubular $MoCo_2O_4$ electrode resulting in an observed decrease in capacitance with the increased discharge current density. This implies ion penetration is feasible at lower current densities where ions have access to the inner structure, and thus all active area of the electrode. However, at higher current densities, the effective use of the material is limited to only the outer surface of the electrode. The specific capacitance of MCo_2O_4 electrodes in this study is compared with the literature values at current density 1 A/g, 2 A/g, and 5 A/g and are listed in Table 4. It is evident from Table 4 that electrochemical performance of bio-templated MCo_2O_4 comparable and, in some cases, outperformed electrodes prepared via other techniques.

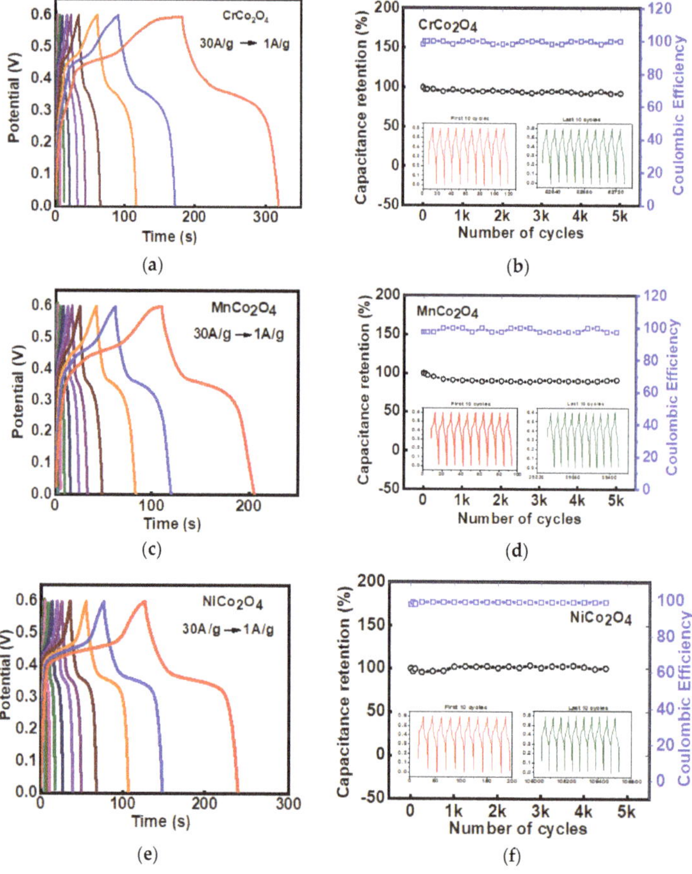

Figure 6. (**a**,**c**,**e**) show charge-discharge (CD) curves of tubular MCo_2O_4 (M = Cr, Mn, Ni) electrode measured in the current density window of 1 to 30 A/g in 3M KOH electrolyte, where red color CD curve is for 1A/g, blue color CD curve is for 1.5A/g, orange color CD curve is for 2A/g and continuously time is decreasing with increasing current density. (**b**,**d**,**f**) show cyclic stability (black color) and coulombic efficiency (blue color) tested at 10 A/g current density for 5000 cycles in 3M KOH electrolytes of MCo_2O_4 (M = Cr, Mn, Ni).

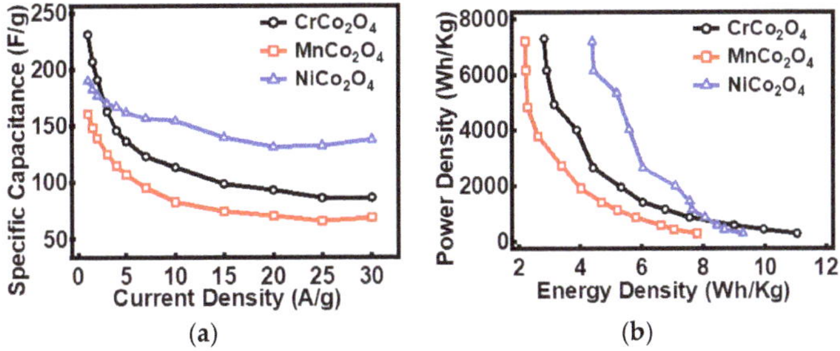

Figure 7. (**a**) Comparison of specific capacitance as a function of current density and (**b**) Ragone plot of power density vs. energy density.

Table 4. Comparison of electrochemical performance of MCo_2O_4 (M = Cr, Mn, Ni) as available from the literature.

Electrode Material	Electrolyte	Specific Capacitance	Energy Density	Power Density	Cyclic Performance (retention)	Ref.
$MnCo_2O_4$ nanofiber		108 F/g at 10 A/g	54 Wh/Kg	9851 W/Kg	85 % after 3000 cycles	[5]
Nanorods $MnCo_2O_4$	1M KOH	349.8 F/g at 1 A/g	35.4 Wh/Kg	225 W/Kg	92.7% after 50 cycles	[86]
Nanoneedles $MnCo_2O_4$	6M KOH	1535 F/g at 1 A/g	35.4 Wh/Kg	225 W/Kg	94.3% after 12000 cycles	[87]
Nanorods $MnCo_2O_4$	2M KOH	845.6 F/g at 1 A/g	35.4 Wh/Kg	225 W/Kg	90.2% after 2000 cycles	[24]
Nanorods $MnCo_2O_4$	1M KOH	308.3 F/g at 1 A/g	55.5 Wh/Kg	5400 W/Kg	88.76% after 2000 cycles	[88]
$MnCo_2O_4$ nanowires @MnO_2	-	2262 F/g at 1 A/g	85.7 Wh/Kg	800 W/Kg	-	[89]
Nanorods $MnCo_2O_4$		718 F/g at 0.5 A/g	-	-	84 % after 1000 cycles	[90]
$NiCo_2O_4$ nanorods	2M KOH	565 F/g at 1 A/g	-	-	77.6% after 1000 cycles	[69]
RGO decorated nanorods bundle $NiCo_2O_4$	6M KOH	1278F/g at 1 A/g	-	-	95% after 1000 cycles	[84]
Nanorods assemble $NiCo_2O_4$	2M KOH	764 F/g at 2 A/g	-	-	101.7% after 1500 cycles	[91]
Nanorods $NiCo_2O_4$	2M KOH	823 F/g at 0.823 A/g	28.51 Wh/Kg	-	101.7% after 1500 cycles	[2]
GO/ Nanorods $NiCo_2O_4$	1M KOH	709.7 F/g at 1 A/g	28 Wh/Kg	8000 W/Kg	94.3% after 5000 cycles	[92]
Nanorods $NiCo_2O_4$	2M KOH	600 F/g at 5 A/g	-	-	80% after 1500 cycles	[70]
Nanorods $NiCo_2O_4$ @PANI	1M H_2So_4	901 F/g at 5 A/g	-	-	91% after 3000 cycles	[93]
Templated $CrCo_2O_4$, $MnCo_2O_4$, and $NiCo_2O_4$ microstructure	3M KOH	403.4, 378, and 407.2 F/g at 2mV/s and 231, 161, and 190 F/g at 1 A/g, respectively	11.1, 7.8, and 9.3 Wh/kg, respectively	7287.3, 7195.3, and 7186.1 W/Kg, respectively	92% after 5000 cycles 91% after 5000 cycles and 100% after 5000 cycles, respectively.	[This work]

Estimation of the electrochemical utilization of the active materials ($CrCo_2O_4$, $MnCo_2O_4$, and $NiCo_2O_4$ electrode), was evaluated from the fraction of cobalt sites, z. The fraction, z, can be evaluated using Faraday's law as following [94]:

$$z = C_{sp}\ MW\ \Delta V/F \quad (12)$$

where C_{sp}, MW, ΔV *and* F is the specific capacitance value, the molecular weight, the applied potential window, and the Faraday's constant, respectively. The z value of 1 indicates the complete involvement of electroactive material. i.e., all active metal sites participating in the redox process. The molecular weight of Co (84.03 g/mol) and Cr (51.996 g/mol) in $CrCo_2O_4$, Co (84.03 g/mol) and Mn (54.93 g/mol) in $MnCo_2O_4$, and Co (84.03 g/mol) and Ni (58.69 g/mol) in $NiCo_2O_4$ the specific capacitance at a current density of 1 A/g (Csp~ 231 F/g, 161 F/g, and 190 F/g) for $CrCo_2O_4$, $MnCo_2O_4$, and $NiCo_2O_4$, Figure 5b, and a potential window of 0.6 V gives a z value of 0.195, 0.139, and 0.169 respectively. In other words, ~20%, 14%, and 17% of the total active material Cr, Mn, and Ni atoms participate in the redox reaction for the charge storage. The observed low value of z suggests that the charge storage in tubular MCo_2O_4 structure via a redox reaction process occurs mainly at the surface with little bulk interaction due to diffusion of OH^- ions into the material. It could be concluded that the charge storage due to the redox process in MCo_2O_4 mostly occurs only at the redox sites predominantly located on the surface of the particles [71].

Cyclic stability tests were performed to evaluate the practical performance of electrodes as a supercapacitor. The stability test of electrode materials was assessed via galvanostatic CD measurement for 5000 cycles for a current density of 10 A g^{-1} in 3M KOH and is shown in Figure 6b,d, and Figure 6f. The Coulombic efficiency (η) of the devices was calculated from its charging (T_c) and the discharging (T_d) times from GCD curves following the relation, $\eta = T_d/T_c \times 100$, and is plotted in Figure 6b,d,f as a function of cyclic time. The initial η of the device was ~100%, which remained practically the same even after 5000 cycles. For the practical applications, the study of cycling performance for electroactive material is very significant parameter. The percentage retention in specific capacitance was calculated using,

$$\%\ \text{retention in specific capacitance} = (C_{\#}/C_1)\times 100 \quad (13)$$

where $C_{\#}$ and C_1 are specific capacitance at various cycles and the 1st cycle, respectively. The specific capacitance of the electrode is reduced by 7.68% in $CrCo_2O_4$, reduced by 9.12% in $MnCo_2O_4$, and increased by 0.48% in $NiCo_2O_4$.

Figure 5h shows the Ragone plots of as-synthesized MCo_2O_4 electrodes. The energy densities (E) and power densities (P) of the electrochemical cells are calculated using the following equations [95]:

$$E = (1/2)CV^2 \quad (14)$$

$$P = E/t \quad (15)$$

where C, V, and t are the specific capacitance that depends on the mass of the electrodes, the operating voltage of the cell, and discharge time in seconds, respectively. The essential point for high-performance supercapacitors is to obtain a high energy density and meanwhile providing an outstanding power density. It is observed from Figure 7b that the tubular $CrCo_2O_4$, $MnCo_2O_4$, and $NiCo_2O_4$ electrode display superior performance over energy density up to 11.1 Wh/kg, 7.8 Wh/kg, and 9.3 Wh/kg with a peak power density up to 7287.34 W/kg, 7195.33 W/kg, and 7186.12 W/kg, respectively are given in Table 3. As supercapacitor is expected to provide higher power and energy density at the same time, hence $NiCo_2O_4$ displays an overall better energy density of 9.3 Wh/kg and a power density of 7186.12 W/kg as compared to $MnCo_2O_4$ and $CrCo_2O_4$.

Ideally, to develop a higher energy density battery with a given anode, a cathode with high electrochemical interaction potential is desired. This is because the energy density of the device equals the product of the working voltage, which is obtained from the electrochemical potentials different between the cathode and anode and specific capacity of the electrode materials [63]. On the other hand, the theoretical capacity of electrode materials depend on the number of reactive electrons (n) and molar weight (M) of the materials and is expressed as Equation (16) [96],

$$C_t = n*F/3.6*M \tag{16}$$

Here, F is the Faraday constant, n is the number of reactive electrons, and M is the molar weight of materials. Theoretically, the equation predicts that an electrode material having a smaller molecular weight can produce a higher capacity. The molecular weight of $CrCo_2O_4$, $MnCo_2O_4$, and $NiCo_2O_4$ is ~ 233.86, 236.80, and 240.55 g/mol, respectively. Thus, in line with the Equation (11), at low current density, among three electrodes studied, the $CrCo_2O_4$ displays higher specific capacitance. However, overall superior performance in terms of energy and power density was displayed by the $NiCo_2O_4$ electrode. $NiCo_2O_4$ is known to possess rich electroactive sites, narrow pore size, and higher electrical conductivity (at least two magnitudes higher) than that of Co_3O_4 and NiO, which could be the reason for the observed overall better performance of $NiCo_2O_4$ [97,98].

4. Conclusions

In conclusion, biomorphic tubular $CrCo_2O_4$, $MnCo_2O_4$, and $NiCo_2O_4$ nanostructures were prepared using cotton by a cost-effective and straightforward bio-template method. The synthesized tubular MCo_2O_4 display excellent crystallinity, phase purity and display desirable electrochemical properties, which indicate a good chance for the fabrication of high-performance supercapacitor devices. Electrodes constructed using the tubular MCo_2O_4 demonstrate high specific capacitance, cyclic stability, power, and energy density when evaluated in 3M KOH electrolyte. The study suggests that it is imperative to account for the nature of the electroactive sites and the conductivity of materials when choosing materials from the series of transition metal oxide as the electrode for the supercapacitor application. Furthermore, the superior electrochemical performance of tubular MCo_2O_4 microstructure owes to the presence of conducting a carbonaceous structure. The highly porous carbonaceous structure can allow electrolyte access throughout the electrode structure. Thus, it produces a large surface area for ion transfer between the electrolyte and the active materials, which leads to achieving ultrafast storage and release of energy.

Author Contributions: Conceptualization, D.G.; methodology, D.G., and C.Z.; software, D.G. and C.Z.; validation, S.R.M., R.K.G., and D.G.; formal analysis, D.G.; investigation, S.R.M. and R.K.G.; resources S.R.M.; data curation, D.G., C.Z., and R.K.G.; writing—original draft preparation, D.G.; writing—review and editing, S.R.M.; visualization, S.R.M. and D.G.; supervision, S.R.M. and R.K.G.; project administration, S.R.M.; funding acquisition, S.R.M. All authors have read and agreed to the published version of the manuscript.

Funding: This is supported by the grants from FIT-DRONES and Biologistics at the University of Memphis, Memphis; TN. Dr. Ram K. Gupta expresses his sincere acknowledgment of the Polymer Chemistry Initiative at Pittsburg State University for providing financial and research support.

Conflicts of Interest: The authors declare no conflict of interest.

References

1. Suktha, P.; Chiochan, P.; Iamprasertkun, P.; Wutthiprom, J.; Phattharasupakun, N.; Suksomboon, M.; Kaewsongpol, T.; Sirisinudomkit, P.; Pettong, T.; Sawangphruk, M. High-performance supercapacitor of functionalized carbon fiber paper with high surface ionic and bulk electronic conductivity: Effect of organic functional groups. *Electrochim. Acta* **2015**, *176*, 504–513. [CrossRef]
2. Sahoo, S.; Ratha, S.; Rout, C.S. Spinel $NiCo_2O_4$ nanorods for supercapacitor applications. *Am. J. Eng. Appl. Sci.* **2015**, *8*, 371–379. [CrossRef]

3. El-Kady, M.F.; Strong, V.; Dubin, S.; Kaner, R.B. Laser scribing of high-performance and flexible graphene-based electrochemical capacitors. *Science* **2012**, *335*, 1326–1330. [CrossRef]
4. Zhu, Y.; Murali, S.; Stoller, M.D.; Ganesh, K.J.; Cai, W.; Ferreira, P.J.; Pirkle, A.; Wallace, R.M.; Cychosz, K.A.; Thommes, M.; et al. Carbon-based supercapacitors produced by activation of graphene. *Science* **2011**, *332*, 1537–1541. [CrossRef]
5. Pettong, T.; Iamprasertkun, P.; Krittayavathananon, A.; Sukha, P.; Sirisinudomkit, P.; Seubsai, A.; Chareonpanich, M.; Kongkachuichay, P.; Limtrakul, J.; Sawangphruk, M. High-performance asymmetric supercapacitors of $MnCo_2O_4$ nanofibers and N-doped reduced graphene oxide aerogel. *ACS Appl. Mater. Interfaces* **2016**, *8*, 34045–34053. [CrossRef] [PubMed]
6. Fan, L.Z.; Qiao, S.; Song, W.; Wu, M.; He, X.; Qu, X. Effects of the functional groups on the electrochemical properties of ordered porous carbon for supercapacitors. *Electrochim. Acta* **2013**, *105*, 299–304. [CrossRef]
7. Chen, X.L.; Li, W.S.; Tan, C.L.; Li, W.; Wu, Y.Z. Improvement in electrochemical capacitance of carbon materials by nitric acid treatment. *J. Power Sources* **2008**, *184*, 668–674. [CrossRef]
8. Yang, Q.; Lu, Z.; Li, T.; Sun, X.; Liu, J. Hierarchical construction of core–shell metal oxide nanoarrays with ultrahigh areal capacitance. *Nano Energy* **2014**, *7*, 170–178. [CrossRef]
9. Vilatela, J.J.; Eder, D. Nanocarbon composites and hybrids in sustainability: A review. *ChemSusChem* **2012**, *5*, 456–478. [CrossRef]
10. Nguyen, T.; Montemor, M.D.F. Metal Oxide and Hydroxide–Based Aqueous Supercapacitors: From Charge Storage Mechanisms and Functional Electrode Engineering to Need-Tailored Devices. *Adv. Sci.* **2019**, *6*, 1801797. [CrossRef]
11. Alqahtani, D.M.; Zequine, C.; Ranaweera, C.K.; Siam, K.; Kahol, P.K.; Poudel, T.P.; Mishra, S.R.; Gupta, R.K. Effect of metal ion substitution on electrochemical properties of cobalt oxide. *J. Alloy. Compd.* **2019**, *771*, 951–959. [CrossRef]
12. Pendashteh, A.; Moosavifard, S.E.; Rahmanifar, M.S.; Wang, Y.; El-Kady, M.F.; Kaner, R.B.; Mousavi, M.F. Highly ordered mesoporous $CuCo_2O_4$ nanowires, a promising solution for high-performance supercapacitors. *Chem. Mater.* **2015**, *27*, 3919–3926. [CrossRef]
13. Cheng, H.; Lu, Z.G.; Deng, J.Q.; Chung, C.Y.; Zhang, K.; Li, Y.Y. A facile method to improve the high rate capability of Co_3O_4 nanowire array electrodes. *Nano Res.* **2010**, *3*, 895–901. [CrossRef]
14. Yan, D.; Zhang, H.; Chen, L.; Zhu, G.; Li, S.; Xu, H.; Yu, A. Biomorphic synthesis of mesoporous Co_3O_4 microtubules and their pseudocapacitive performance. *ACS Appl. Mater. Interfaces* **2014**, *6*, 15632–15637. [CrossRef] [PubMed]
15. Sieber, H. Biomimetic synthesis of ceramics and ceramic composites. *Mater. Sci. Eng. A* **2005**, *412*, 43–47. [CrossRef]
16. Habibi, Y.; Lucia, L.A.; Rojas, O.J. Cellulose nanocrystals: Chemistry, self-assembly, and applications. *Chem. Rev.* **2010**, *110*, 3479–3500. [CrossRef]
17. Liu, Y.; Lv, B.; Li, P.; Chen, Y.; Gao, B.; Lin, B. Biotemplate-assisted hydrothermal synthesis of tubular porous Co_3O_4 with excellent charge-discharge cycle stability for supercapacitive electrodes. *Mater. Lett.* **2018**, *210*, 231–234. [CrossRef]
18. Fan, T.X.; Chow, S.K.; Zhang, D. Biomorphic mineralization: From biology to materials. *Prog. Mater. Sci.* **2009**, *54*, 542–659. [CrossRef]
19. Sotiropoulou, S.; Sierra-Sastre, Y.; Mark, S.S.; Batt, C.A. Biotemplated nanostructured materials. *Chem. Mater.* **2008**, *20*, 821–834. [CrossRef]
20. Zhou, H.; Fan, T.; Zhang, D. Biotemplated materials for sustainable energy and environment: Current status and challenges. *ChemSusChem* **2011**, *4*, 1344–1387. [CrossRef]
21. Shim, H.W.; Lim, A.H.; Kim, J.C.; Jang, E.; Seo, S.D.; Lee, G.H.; Kim, T.D.; Kim, D.W. Scalable one-pot bacteria-templating synthesis route toward hierarchical, porous-Co_3O_4 superstructures for supercapacitor electrodes. *Sci. Rep.* **2013**, *3*, 2325. [CrossRef] [PubMed]
22. Han, L.; Yang, D.P.; Liu, A. Leaf-templated synthesis of 3D hierarchical porous cobalt oxide nanostructure as direct electrochemical biosensing interface with enhanced electrocatalysis. *Biosens. Bioelectron.* **2015**, *63*, 145–152. [CrossRef]
23. Song, P.; Zhang, H.; Han, D.; Li, J.; Yang, Z.; Wang, Q. Preparation of biomorphic porous $LaFeO_3$ by sorghum straw biotemplate method and its acetone sensing properties. *Sens. Actuators B Chem.* **2014**, *196*, 140–146. [CrossRef]

24. Weatherspoon, M.R.; Cai, Y.; Crne, M.; Srinivasarao, M.; Sandhage, K.H. 3D Rutile Titania-Based Structures with Morpho Butterfly Wing Scale Morphologies. *Angew. Chem. Int. Ed.* **2008**, *47*, 7921–7923. [CrossRef] [PubMed]
25. Yan, D.; Li, S.; Zhu, G.; Wang, Z.; Xu, H.; Yu, A. Synthesis and pseudocapacitive behaviors of biomorphic mesoporous tubular MnO_2 templated from cotton. *Mater. Lett.* **2013**, *95*, 164–167. [CrossRef]
26. Mondal, A.K.; Su, D.; Chen, S.; Ung, A.; Kim, H.S.; Wang, G. Mesoporous $MnCo_2O_4$ with a flake-like structure as advanced electrode materials for lithium-ion batteries and supercapacitors. *Chem. A Eur. J.* **2015**, *21*, 1526–1532. [CrossRef]
27. Xu, J.; Sun, Y.; Lu, M.; Wang, L.; Zhang, J.; Tao, E.; Qian, J.; Liu, X. Fabrication of the porous $MnCo_2O_4$ nanorod arrays on Ni foam as an advanced electrode for asymmetric supercapacitors. *Acta Mater.* **2018**, *152*, 162–174. [CrossRef]
28. Dong, Y.; Wang, Y.; Xu, Y.; Chen, C.; Wang, Y.; Jiao, L.; Yuan, H. Facile synthesis of hierarchical nanocage $MnCo_2O_4$ for high performance supercapacitor. *Electrochim. Acta* **2017**, *225*, 39–46. [CrossRef]
29. Yang, J.; Cho, M.; Lee, Y. Synthesis of hierarchical $NiCo_2O_4$ hollow nanorods via sacrificial-template accelerate hydrolysis for electrochemical glucose oxidation. *Biosens. Bioelectron.* **2016**, *75*, 15–22. [CrossRef]
30. Zhao, Z.; Geng, F.; Bai, J.; Cheng, H.M. Facile and controlled synthesis of 3D nanorods-based urchinlike and nanosheets-based flowerlike cobalt basic salt nanostructures. *J. Phys. Chem. C* **2007**, *111*, 3848–3852. [CrossRef]
31. Jadhav, A.R.; Bandal, H.A.; Kim, H. $NiCo_2O_4$ hollow sphere as an efficient catalyst for hydrogen generation by NaBH4 hydrolysis. *Mater. Lett.* **2017**, *198*, 50–53. [CrossRef]
32. Huang, W.; Lin, T.; Cao, Y.; Lai, X.; Peng, J.; Tu, J. Hierarchical $NiCo_2O_4$ hollow sphere as a peroxidase mimetic for colorimetric detection of H_2O_2 and glucose. *Sensors* **2017**, *17*, 217. [CrossRef]
33. Pu, J.; Wang, J.; Jin, X.; Cui, F.; Sheng, E.; Wang, Z. Porous hexagonal $NiCo_2O_4$ nanoplates as electrode materials for supercapacitors. *Electrochim. Acta* **2013**, *106*, 226–234. [CrossRef]
34. Park, J.; Shen, X.; Wang, G. Solvothermal synthesis and gas-sensing performance of Co_3O_4 hollow nanospheres. *Sens. Actuators B Chem.* **2009**, *136*, 494–498. [CrossRef]
35. Lasheras, X.; Insausti, M.; Gil de Muro, I.; Garaio, E.; Plazaola, F.; Moros, M.; De Matteis, L.; M de la Fuente, J.; Lezama, L. Chemical synthesis and magnetic properties of monodisperse nickel ferrite nanoparticles for biomedical applications. *J. Phys. Chem. C* **2016**, *120*, 3492–3500. [CrossRef]
36. Kuang, L.; Ji, F.; Pan, X.; Wang, D.; Chen, X.; Jiang, D.; Zhang, Y.; Ding, B. Mesoporous $MnCo_2O_4$ nanoneedle arrays electrode for high-performance asymmetric supercapacitor application. *Chem. Eng. J.* **2017**, *315*, 491–499. [CrossRef]
37. Melot, B.C.; Tarascon, J.M. Design and preparation of materials for advanced electrochemical storage. *Acc. Chem. Res.* **2013**, *46*, 1226–1238. [CrossRef]
38. Chen, K.; Xue, D. Materials chemistry toward electrochemical energy storage. *J. Mater. Chem. A* **2016**, *4*, 7522–7537. [CrossRef]
39. *Principles and Applications of Lithium Secondary Batteries*; Park, J.K. (Ed.) John Wiley & Sons: Hoboken, NJ, USA, 2012.
40. Zhang, Y.; Ma, M.; Yang, J.; Su, H.; Huang, W.; Dong, X. Selective synthesis of hierarchical mesoporous spinel $NiCo_2O_4$ for high-performance supercapacitors. *Nanoscale* **2014**, *6*, 4303–4308. [CrossRef]
41. Yuan, C.; Wu, H.B.; Xie, Y.; Lou, X.W. Mixed transition-metal oxides: Design, synthesis, and energy-related applications. *Angew. Chem. Int. Ed.* **2014**, *53*, 1488–1504. [CrossRef]
42. Zhang, L.L.; Zhao, X.S. Carbon-based materials as supercapacitor electrodes. *Chem. Soc. Rev.* **2009**, *38*, 2520–2531. [CrossRef]
43. Zheng, Y.Z.; Ding, H.; Uchaker, E.; Tao, X.; Chen, J.F.; Zhang, Q.; Cao, G. Nickel-mediated polyol synthesis of hierarchical V_2O_5 hollow microspheres with enhanced lithium storage properties. *J. Mater. Chem. A* **2015**, *3*, 1979–1985. [CrossRef]
44. *Lithium Batteries: Science and Technology*; Nazri, G.A.; Pistoia, G. (Eds.) Springer Science & Business Media: Berlin/Heidelberg, Germany, 2008.
45. Li, J.; Wang, J.; Liang, X.; Zhang, Z.; Liu, H.; Qian, Y.; Xiong, S. Hollow $MnCo_2O_4$ submicrospheres with multilevel interiors: From mesoporous spheres to yolk-in-double-shell structures. *ACS Appl. Mater. Interfaces* **2013**, *6*, 24–30. [CrossRef] [PubMed]

46. Bhojane, P.; Sen, S.; Shirage, P.M. Enhanced electrochemical performance of mesoporous $NiCo_2O_4$ as an excellent supercapacitive alternative energy storage material. *Appl. Surf. Sci.* **2016**, *377*, 376–384. [CrossRef]
47. Jenkins, R.; Snyder, R.L. *Introduction to X-ray Powder Diffractometry*; (No. 543.427 JEN); Wiley: Hoboken, NJ, USA, 1996.
48. Lin, H.K.; Chiu, H.C.; Tsai, H.C.; Chien, S.H.; Wang, C.B. Synthesis, characterization and catalytic oxidation of carbon monoxide over cobalt oxide. *Catal. Lett.* **2003**, *88*, 169–174. [CrossRef]
49. Spencer, C.D.; Schroeer, D. Mössbauer study of several cobalt spinels using Co^{57} and Fe^{57}. *Phys. Rev. B* **1974**, *9*, 3658. [CrossRef]
50. Kurtulus, F.; Guler, H. A Simple Microwave-Assisted Route to Prepare Black Cobalt, Co_3O_4. *Inorg. Mater.* **2005**, *41*, 483–485. [CrossRef]
51. St, G.C.; Stoyanova, M.; Georgieva, M.; Mehandjiev, D. Preparation and characterization of a higher cobalt oxide. *Mater. Chem. Phys.* **1999**, *60*, 39–43.
52. Khairy, M.; Mousa, M. Synthesis of Ternary and Quaternary metal oxides based on Ni, Mn, Cu, and Co for high-performance Supercapacitor. *J. Ovonic Res.* **2019**, *15*, 181–198.
53. Štěpánek, F.; Marek, M.; Adler, P.M. Modeling capillary condensation hysteresis cycles in reconstructed porous media. *Aiche J.* **1999**, *45*, 1901–1912. [CrossRef]
54. Wang, R.; Li, Q.; Cheng, L.; Li, H.; Wang, B.; Zhao, X.S.; Guo, P. Electrochemical properties of manganese ferrite-based supercapacitors in aqueous electrolyte: The effect of ionic radius. *Colloid. Surfaces A Physicochem. Eng. Aspects* **2014**, *457*, 94–99. [CrossRef]
55. Hou, L.; Yuan, C.; Yang, L.; Shen, L.; Zhang, F.; Zhang, X. Urchin-like Co_3O_4 microspherical hierarchical superstructures constructed by one-dimension nanowires toward electrochemical capacitors. *Rsc Adv.* **2011**, *1*, 1521–1526. [CrossRef]
56. Yuan, C.; Yang, L.; Hou, L.; Shen, L.; Zhang, X.; Lou, X.W.D. Growth of ultrathin mesoporous Co_3O_4 nanosheet arrays on Ni foam for high-performance electrochemical capacitors. *Energy Environ. Sci.* **2012**, *5*, 7883–7887. [CrossRef]
57. Adhikari, H.; Ghimire, M.; Ranaweera, C.K.; Bhoyate, S.; Gupta, R.K.; Alam, J.; Mishra, S.R. Synthesis and electrochemical performance of hydrothermally synthesized Co_3O_4 nanostructured particles in presence of urea. *J. Alloy. Compd.* **2017**, *708*, 628–638. [CrossRef]
58. Sun, D.; He, L.; Chen, R.; Liu, Y.; Lv, B.; Lin, S.; Lin, B. Biomorphic composites composed of octahedral Co3O4 nanocrystals and mesoporous carbon microtubes templated from cotton for excellent supercapacitor electrodes. *Appl. Surf. Sci.* **2019**, *465*, 232–240. [CrossRef]
59. Brousse, T.; Bélanger, D. A Hybrid Fe_3O_4 MnO_2 Capacitor in Mild Aqueous Electrolyte. *Electrochem. Solid-State Lett.* **2003**, *6*, A244–A248. [CrossRef]
60. Wang, S.Y.; Ho, K.C.; Kuo, S.L.; Wu, N.L. Investigation on capacitance mechanisms of Fe_3O_4 electrochemical capacitors. *J. Electrochem. Soc.* **2006**, *153*, A75–A80. [CrossRef]
61. Tiruye, G.A.; Munoz-Torrero, D.; Palma, J.; Anderson, M.; Marcilla, R. All-solid state supercapacitors operating at 3.5 V by using ionic liquid based polymer electrolytes. *J. Power Sources* **2015**, *279*, 472–480. [CrossRef]
62. Gao, F.; Shao, G.; Qu, J.; Lv, S.; Li, Y.; Wu, M. Tailoring of porous and nitrogen-rich carbons derived from hydrochar for high-performance supercapacitor electrodes. *Electrochim. Acta* **2015**, *155*, 201–208. [CrossRef]
63. Selvam, M.; Srither, S.R.; Saminathan, K.; Rajendran, V. Chemically and electrochemically prepared graphene/MnO_2 nanocomposite electrodes for zinc primary cells: A comparative study. *Ionics* **2015**, *21*, 791–799. [CrossRef]
64. Tang, Y.; Liu, Y.; Yu, S.; Gao, F.; Zhao, Y. Comparative study on three commercial carbons for supercapacitor applications. *Russ. J. Electrochem.* **2015**, *51*, 77–85. [CrossRef]
65. Sankar, K.V.; Selvan, R.K. Improved electrochemical performances of reduced graphene oxide based supercapacitor using redox additive electrolyte. *Carbon* **2015**, *90*, 260–273. [CrossRef]
66. Sahu, V.; Shekhar, S.; Sharma, R.K.; Singh, G. Ultrahigh performance supercapacitor from lacey reduced graphene oxide nanoribbons. *ACS Appl. Mater. Interfaces* **2015**, *7*, 3110–3116. [CrossRef] [PubMed]
67. Fic, K.; Platek, A.; Piwek, J.; Frackowiak, E. Sustainable materials for electrochemical capacitors. *Mater. Today* **2018**, *21*, 437–454. [CrossRef]
68. Meher, S.K.; Justin, P.; Rao, G.R. Nanoscale morphology dependent pseudocapacitance of NiO: Influence of intercalating anions during synthesis. *Nanoscale* **2011**, *3*, 683–692. [CrossRef]

69. Xiao, J.; Yang, S. Sequential crystallization of sea urchin-like bimetallic (Ni, Co) carbonate hydroxide and its morphology conserved conversion to porous $NiCo_2O_4$ spinel for pseudocapacitors. *Rsc Adv.* **2011**, *1*, 588–595. [CrossRef]
70. Liu, X.Y.; Zhang, Y.Q.; Xia, X.H.; Shi, S.J.; Lu, Y.; Wang, X.L.; Gu, C.D.; Tu, J.P. Self-assembled porous $NiCo_2O_4$ hetero-structure array for electrochemical capacitor. *J. Power Sources* **2013**, *239*, 157–163. [CrossRef]
71. Wang, X.; Han, X.; Lim, M.; Singh, N.; Gan, C.L.; Jan, M.; Lee, P.S. Nickel cobalt oxide-single wall carbon nanotube composite material for superior cycling stability and high-performance supercapacitor application. *J. Phys. Chem. C* **2012**, *116*, 12448–12454. [CrossRef]
72. Liu, X.; Long, Q.; Jiang, C.; Zhan, B.; Li, C.; Liu, S.; Zhao, Q.; Huang, W.; Dong, X. Facile and green synthesis of mesoporous Co_3O_4 nanocubes and their applications for supercapacitors. *Nanoscale* **2013**, *5*, 6525–6529. [CrossRef]
73. Zhu, Y.; Pu, X.; Song, W.; Wu, Z.; Zhou, Z.; He, X.; Lu, F.; Jing, M.; Tang, B.; Ji, X. High capacity $NiCo_2O_4$ nanorods as electrode materials for supercapacitor. *J. Alloy. Compd.* **2014**, *617*, 988–993. [CrossRef]
74. Jokar, E.; Shahrokhian, S. Synthesis and characterization of $NiCo_2O_4$ nanorods for preparation of supercapacitor electrodes. *J. Solid State Electrochem.* **2015**, *19*, 269–274. [CrossRef]
75. Veeramani, V.; Madhu, R.; Chen, S.M.; Sivakumar, M.; Hung, C.T.; Miyamoto, N.; Liu, S.B. $NiCo_2O_4$-decorated porous carbon nanosheets for high-performance supercapacitors. *Electrochim. Acta* **2017**, *247*, 288–295. [CrossRef]
76. Zhai, Y.; Mao, H.; Liu, P.; Ren, X.; Xu, L.; Qian, Y. Facile fabrication of hierarchical porous rose-like $NiCo_2O_4$ nanoflake/$MnCo_2O_4$ nanoparticle composites with enhanced electrochemical performance for energy storage. *J. Mater. Chem. A* **2015**, *3*, 16142–16149. [CrossRef]
77. Ghosh, D.; Giri, S.; Das, C.K. Hydrothermal synthesis of platelet β $Co(OH)_2$ and Co_3O_4: Smart electrode material for energy storage application. *Environ. Prog. Sustain. Energy* **2014**, *33*, 1059–1064. [CrossRef]
78. Yi, H.; Wang, H.; Jing, Y.; Peng, T.; Wang, X. Asymmetric supercapacitors based on carbon nanotubes@ NiO ultrathin nanosheets core-shell composites and MOF-derived porous carbon polyhedrons with super-long cycle life. *J. Power Sources* **2015**, *285*, 281–290. [CrossRef]
79. Shen, B.; Guo, R.; Lang, J.; Liu, L.; Liu, L.; Yan, X. A high-temperature flexible supercapacitor based on pseudocapacitive behavior of FeOOH in an ionic liquid electrolyte. *J. Mater. Chem. A* **2016**, *4*, 8316–8327. [CrossRef]
80. Augustyn, V.; Come, J.; Lowe, M.A.; Kim, J.W.; Taberna, P.L.; Tolbert, S.H.; Abruña, H.D.; Simon, P.; Dunn, B. High-rate electrochemical energy storage through Li^+ intercalation pseudocapacitance. *Nat. Mater.* **2013**, *12*, 518. [CrossRef]
81. Lindström, H.; Södergren, S.; Solbrand, A.; Rensmo, H.; Hjelm, J.; Hagfeldt, A.; Lindquist, S.E. Li^+ ion insertion in TiO_2 (anatase). 1. Chronoamperometry on CVD films and nanoporous films. *J. Phys. Chem. B* **1997**, *101*, 7710–7716. [CrossRef]
82. Liu, T.C.; Pell, W.G.; Conway, B.E.; Roberson, S.L. Behavior of molybdenum nitrides as materials for electrochemical capacitors comparison with ruthenium oxide. *J. Electrochem. Soc.* **1998**, *145*, 1882–1888. [CrossRef]
83. Wang, J.; Polleux, J.; Lim, J.; Dunn, B. Pseudocapacitive contributions to electrochemical energy storage in TiO_2 (anatase) nanoparticles. *J. Phys. Chem. C* **2007**, *111*, 14925–14931. [CrossRef]
84. Zhao, D.D.; Bao, S.J.; Zhou, W.J.; Li, H.L. Preparation of hexagonal nanoporous nickel hydroxide film and its application for electrochemical capacitor. *Electrochem. Commun.* **2007**, *9*, 869–874. [CrossRef]
85. Sugimoto, W.; Iwata, H.; Yasunaga, Y.; Murakami, Y.; Takasu, Y. Preparation of ruthenic acid nanosheets and utilization of its interlayer surface for electrochemical energy storage. *Angew. Chem. Int. Ed.* **2003**, *42*, 4092–4096. [CrossRef] [PubMed]
86. Li, L.; Zhang, Y.Q.; Liu, X.Y.; Shi, S.J.; Zhao, X.Y.; Zhang, H.; Ge, X.; Cai, G.F.; Gu, C.D.; Wang, X.L.; et al. One-dimension $MnCo_2O_4$ nanowire arrays for electrochemical energy storage. *Electrochim. Acta* **2014**, *116*, 467–474. [CrossRef]
87. Hui, K.N.; San Hui, K.; Tang, Z.; Jadhav, V.V.; Xia, Q.X. Hierarchical chestnut-like $MnCo_2O_4$ nanoneedles grown on nickel foam as binder-free electrode for high energy density asymmetric supercapacitors. *J. Power Sources* **2016**, *330*, 195–203. [CrossRef]
88. Shi, X.; Liu, Z.; Zheng, Y.; Zhou, G. Controllable synthesis and electrochemical properties of $MnCo_2O_4$ nanorods and microcubes. *Colloids Surf. A: Physicochem. Eng. Asp.* **2017**, *522*, 525–535. [CrossRef]

89. Liu, S.; Hui, K.S.; Hui, K.N. 1 D hierarchical $MnCo_2O_4$ nanowire@ MnO_2 sheet core–shell arrays on graphite paper as superior electrodes for asymmetric supercapacitors. *ChemNanoMat* **2015**, *1*, 593–602. [CrossRef]
90. Venkatachalam, V.; Alsalme, A.; Alghamdi, A.; Jayavel, R. High performance electrochemical capacitor based on $MnCo_2O_4$ nanostructured electrode. *J. Electroanal. Chem.* **2015**, *756*, 94–100. [CrossRef]
91. Zhu, Y.; Ji, X.; Yin, R.; Hu, Z.; Qiu, X.; Wu, Z.; Liu, Y. Nanorod-assembled $NiCo_2O_4$ hollow microspheres assisted by an ionic liquid as advanced electrode materials for supercapacitors. *Rsc Adv.* **2017**, *7*, 11123–11128. [CrossRef]
92. Mao, J.W.; He, C.H.; Qi, J.Q.; Zhang, A.B.; Sui, Y.W.; He, Y.Z.; Meng, Q.K.; Wei, F.X. An Asymmetric Supercapacitor with Mesoporous $NiCo_2O_4$ Nanorod/Graphene Composite and N-Doped Graphene Electrodes. *J. Electron. Mater.* **2018**, *47*, 512–520. [CrossRef]
93. Jabeen, N.; Xia, Q.; Yang, M.; Xia, H. Unique core–shell nanorod arrays with polyaniline deposited into mesoporous $NiCo_2O_4$ support for high-performance supercapacitor electrodes. *Acs Appl. Mater. Interfaces* **2016**, *8*, 6093–6100. [CrossRef]
94. Srinivasan, V.; Weidner, J.W. Studies on the capacitance of nickel oxide films: Effect of heating temperature and electrolyte concentration. *J. Electrochem. Soc.* **2000**, *147*, 880–885. [CrossRef]
95. Lin, C.; Ritter, J.A.; Popov, B.N. Characterization of sol-gel-derived cobalt oxide xerogels as electrochemical capacitors. *J. Electrochem. Soc.* **1998**, *145*, 4097–4103. [CrossRef]
96. Zhi, M.; Xiang, C.; Li, J.; Li, M.; Wu, N. Nanostructured carbon–metal oxide composite electrodes for supercapacitors: A review. *Nanoscale* **2013**, *5*, 72–88. [CrossRef] [PubMed]
97. Wu, Z.; Zhu, Y.; Ji, X. $NiCo_2O_4$-based materials for electrochemical supercapacitors. *J. Mater. Chem. A* **2014**, *2*, 14759–14772. [CrossRef]
98. Bitla, Y.; Chin, Y.Y.; Lin, J.C.; Van, C.N.; Liu, R.; Zhu, Y.; Liu, H.J.; Zhan, Q.; Lin, H.J.; Chen, C.T.; et al. Origin of metallic behavior in NiCo 2 O 4 ferrimagnet. *Sci. Rep.* **2015**, *5*, 15201. [CrossRef]

Article

Investigation and Improvement of Scalable Oxygen Reducing Cathodes for Microbial Fuel Cells by Spray Coating

Thorben Muddemann [1,*,†], Dennis Haupt [2,†], Bolong Jiang [1], Michael Sievers [2] and Ulrich Kunz [1]

1 Institute of Chemical and Electrochemical Process Engineering, Clausthal University of Technology, 38678 Clausthal-Zellerfeld, Germany; bolong.jiang@tu-clausthal.de (B.J.); kunz@icvt.tu-clausthal.de (U.K.)
2 CUTEC Research Center for Environmental Technologies, 38678 Clausthal-Zellerfeld, Germany; dennis.haupt@cutec.de (D.H.); michael.sievers@cutec.de (M.S.)
* Correspondence: thorben.muddemann@tu-clausthal.de
† These authors contributed equally to this work.

Received: 5 November 2019; Accepted: 17 December 2019; Published: 19 December 2019

Abstract: This contribution describes the effect of the quality of the catalyst coating of cathodes for wastewater treatment by microbial fuel cells (MFC). The increase in coating quality led to a strong increase in MFC performance in terms of peak power density and long-term stability. This more uniform coating was realized by an airbrush coating method for applying a self-developed polymeric solution containing different catalysts (MnO_2, MoS_2, Co_3O_4). In addition to the possible automation of the presented coating, this method did not require a calcination step. A cathode coated with catalysts, for instance, MnO_2/MoS_2 (weight ratio 2:1), by airbrush method reached a peak and long-term power density of 320 and 200–240 mW/m^2, respectively, in a two-chamber MFC. The long-term performance was approximately three times higher than a cathode with the same catalyst system but coated with the former paintbrush method on a smaller cathode surface area. This extraordinary increase in MFC performance confirmed the high impact of catalyst coating quality, which could be stronger than variations in catalyst concentration and composition, as well as in cathode surface area.

Keywords: microbial fuel cell; wastewater treatment; oxygen reduction reaction; municipal wastewater; MnO_2; MoS_2; Co_3O_4; spray method

1. Introduction

The permanent rise in standard of living correlates with an increasing energy demand [1]. Particularly in view of global warming and the intended abandonment of fossil and nuclear fuels, processes must be developed that provide environmentally friendly energy [2]. Another challenge is water scarcity, causing 1.2 billion people to suffer from a lack of water and leaving another 1.6 billion people without any access to hygienically safe drinking water [3]. Microbial fuel cells (MFC) are a promising technology that can contribute to a decentralized solution for these future challenges, enabling expansion and assurance of water availability and energy supply [4,5].

Microbial fuel cells (MFCs) combine electrical energy generation with simultaneous wastewater treatment by microorganisms [5]. Anodic exoelectrogenic bacteria oxidize organic compounds from wastewater [6,7]. In combination with a coupled reduction reaction, a positive potential difference, and therefore a net power, is achieved [8]. Oxygen is usually used as final electron acceptor in the coupled oxygen reduction reaction (ORR) at the cathode, as oxygen is available in ambient air [8]. The schematic structure of the investigated MFC is shown in Figure 1.

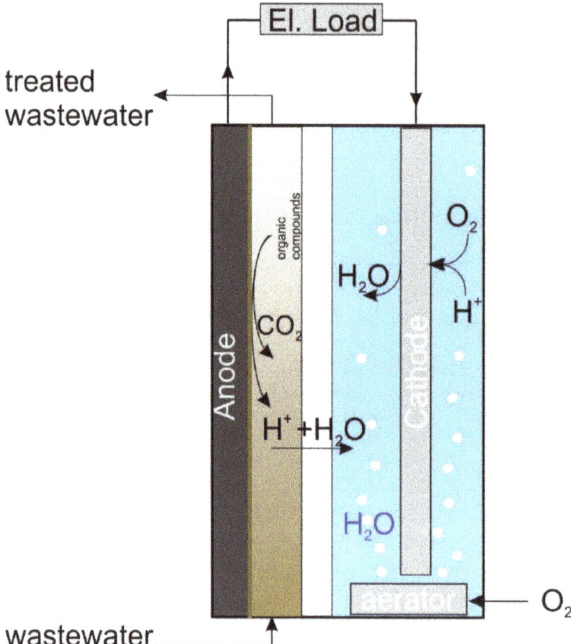

Figure 1. Schematic drawing of the investigated microbial fuel cell (MFC): a planar anode overgrown with bacteria (**dark gray**), an oxygen-reducing cathode (**light gray**), and cation exchange membrane (**white**). The desired reactions and subsequent products are sketched. An electrical load (El. Load) is placed in the outer electrical circuit to use the generated power.

The net power generation of MFCs is rather low, especially due to the performance-limiting cathode [9–12]. To overcome this bottleneck, large electrode surfaces [13], improved cathode manufacturing methods, and especially scalable methods are needed [5]. In the laboratory, cathodes for MFCs are often prepared by brushing [9,14–18], dipping [19], rolling [20], pressing [21], doctor blading [12], spin coating [22], or electro deposition [19,23]. Most of these methods are either not applicable, or less so, for economical technical scale electrodes because they are time-consuming, complex, or not adaptable. Several patents describe manufacturing methods for large-scale cathodes for chemical fuel cells [24–26] or chlor-alkali electrolysis [27,28], but these methods are not yet adapted for MFCs. For that reason, a cathode-coating method is presented in this study, which is also a promising approach for automation, and for the use of large-scale cathodes for MFCs for the first time.

This coating is realized by an airbrush method for applying a self-developed polymeric solution containing catalysts. The coating process was investigated with different catalysts, their different mixing ratios, and different metal meshes.

Next to carbon, MnO_2, MoS_2 (with varying mixing ratios), and Co_3O_4 (with TiO_2 as support material) were investigated as additional catalysts for ORR. As a benchmark for ORR [8], one electrode was coated with Pt as the catalyst. Unfortunately, Pt is very expensive and susceptible to catalyst poisoning by S_2^- or NH_4^+ [29], which are commonly present in municipal wastewater, and as a consequence Pt is not suitable for wastewater treatment applications.

Several publications investigated MnO_2 as an ORR catalyst for MFCs, including the influence of its different modifications (alpha, beta, gamma), different mixing ratios, and varying manufacturing methods. Modifications of MnO_2 were studied by Zhang et al. in an MFC (glucose as feed). The maximum power densities differ from 125 mW/m^2 (alpha) to 172 mW/m^2 (beta) and 88 mW/m^2 (gamma) [30]. This approach was followed by Roche et al. with similar performances up to 161 mW/m^2

(synthetic wastewater) [31]. Newer studies focus on more complex MnO_2-based cathodes. Li et al. studied a manganese oxide catalyst with a cryptomelane-type octahedral molecular sieve structure. This catalyst was additionally doped with Co, Cu, and Ce. The highest power density was measured for the Co-doped system with 180 mW/m^2 (domestic wastewater, acetate) [16], followed by an optimized system with a continuous flow MFC. The maximum power density was 201 mW/m^2 (wastewater inoculum; acetate as feed) [17]. Recent studies partially increased the performance of MnO_2 even further. Zhou et al. investigated the behavior of the combination of polyaniline and MnO_2 nanocomposites and reached power densities of 248 mW/m^2 (anaerobic digester sludge inoculum) [32]. Furthermore, nano-scaled systems based on low-cost MnO_2 nanowires (on carbon support) achieved a maximum of 180 mW/m^2 (combination of domestic and artificial wastewater) [33]. Farahani et al. investigated another MnO_x-based cathode and found that nitrogen-doped carbon with MnO_x revealed a peak power density of 467 mW/m^2 (acetate as feed) [34]. However, the described cathode production process is complex and is difficult to scale. A further study with modified MnO_2 and induced carbon nanotubes (CNT) revealed a drastically lower power density of approximately 12 mW/m^2 (calculated with the given volumetric power density of 216 mW/m^3) [35].

Jiang et al. investigated the performance of MnO_2 and its combination with MoS_2 (real wastewater). The best combination of MnO_2 and MoS_2 revealed peak power densities of up to 165 mW/m^2 [9,14]. Another investigated composition within this study was MoS_2, which is sparsely described in literature. The combination of MoS_2/C is considerably less investigated than MnO_x catalysts. Hao et al. evaluated N-doped MoS_2/C catalysts and reached high power densities (815 mW/m^2) in a small reactor (28 mL) with artificial wastewater (at 30 °C) due to the beneficial catalyst and the optimized process conditions [20].

In comparison, Co_3O_4/C cathodes are more often studied. Ge et al. revealed power densities of up to 1500 mW/m^2 in an MFC (artificial wastewater). This was confirmed by Xia et al. with power densities of up to 1540 mW/m^2. The performance of real wastewater with high chemical oxygen demand (COD; 56,500 mg/L) was investigated by Kumar et al. They reached power densities up to 503 mW/m^2 [36]. A following study with flower-like Co_3O_4 and lower COD loads (digester sludge and acetate) showed lower maximum power densities of 248 mW/m^2 [15]. Our working group already demonstrated the long-term stability of MnO_2 [14], MoS_2 [9], and Co_3O_4 [37] electrodes. It was shown that MnO_2, MoS_2, and Co_3O_4 have decent ORR properties and fulfill the requirements of MFC cathodes—they are inexpensive, long-term stable, environmentally friendly [38,39], reduce the overvoltage, and are immune to catalyst poising. Unfortunately, these cathodes were produced by a time-consuming manual paintbrush method. Therefore, a novel spray coating method newly adapted to MFC applications by a dedicated suspension is presented, which is also adaptable to automation processes and scalable electrodes.

2. Materials and Methods

2.1. Experimantal Setup

To compare and characterize all coated electrodes, a laboratory MFC (self-made) was installed. The test facility consists of eight identical test cells, and an overview is given in Figure 2. Purchased graphite compound anodes were used, made of 86% graphite and 14% olefinic polymer binder (Eisenhuth GmbH, Osterode am Harz, Germany). This material allowed the microorganisms to grow on the surface and was also scalable. A low-cost membrane type FKS-PET 130 (Fumatech GmbH, Bietigheim-Bissingen, Germany) was used to separate the anaerobic anode chamber and the aerobic cathode chamber. For more information regarding the cell and anode design, refer to [14].

Figure 2. Presentation of the used laboratory system, whereby all measurements were done in the white- marked MFC for comparability.

Each MFC was fed with wastewater via a pump (Pumpen und Anlagentechnik, Lutherstadt Wittenberg, Germany, model: 18ZP-VA 0,68-D30-118 FU) and all cathodes were connected to one water basin aerated with ambient air. For all characterizations, the same test cell (Figure 2, circle) was used to increase the measurements' comparability.

During the operation, the temperature, pH values, electrical potentials, cell voltage, and current were monitored through a LabView system (National instruments, Austin, TX, USA). All experiments were carried out at a temperature of approximately 16.5 °C (room temperature in the laboratory). The pH values were kept at approximately 8 (anolyte) and approximately 10 (catholyte) by adding a sodium carbonate solution (20 wt %) (Carl Roth GmbH & Co. KG, Karlsruhe, Germany). A constant current source was used to enable a defined microbial growth at continuously controlled maximum power point conditions. For detailed information about the constant current source, refer to [9,14].

The inoculation process took 2 weeks and was performed with real wastewater taken after the primary clarifier of the sewage treatment plant (STP) Goslar (Eurawasser Betriebsführungsgesellschaft mbH, Goslar, Germany). After filling the anode tank with about 10 L of wastewater and turning on the pumps, the cells were kept in open circuit for 5 days. Afterwards, each cell was connected to its own constant current source for individual cell operation and regulation.

To evaluate the electrodes under real conditions, the effluent of the primary clarifier from the same STP Goslar with an initial COD of 150 to 200 mg/L was treated. Despite weekly wastewater exchanges, 10 mL of a solution of 200 g/L glucose (Carl Roth GmbH & Co. KG, Karlsruhe, Germany). and 200 g/L NaAc (Carl Roth GmbH & Co. KG, Karlsruhe, Germany) was added daily to keep a constant COD concentration level of approximately 200 mg/L, and to ensure the comparability of results at different dates.

2.2. Catalysts and Carrier Materials for Cathode

Printex 6L carbon (Orion Engineered Carbons S.A., Houston, TX, USA) was used for the electrode coating. MnO_2 by EMD Millipore Corporation (article number: 805958) (Burlington, MA, USA) and MoS_2 by Metallpulver24 Corp. (article number: 22020) (Sakt Augustin, Germany) were used as catalysts. Butanone served as solvent (Sigma-Aldrich, St. Louis, MO, USA, 443468). Due to the high prices of Co_3O_4, carrier supported catalysts (Co_3O_4 @ TiO_2) were evaluated. For these Co_3O_4 catalysts, $Co(NO_3)_2$ was obtained by Sigma-Aldrich (article number: 239267) (St. Louis, MO, USA) and anatase TiO_2 nano particles by Cofermin Chemical GmbH (Essen, Germany). Anatase TiO_2 in a whiskers morphology was produced according to reference [40]. Fuel cell grade Pt black by De Nora North America ETEK Division (S990670; Painesville, OH, USA) was used for the Pt coating for benchmarking.

Three different types of metal mesh were used as carrier material for the coating. The physical details of the investigated meshes are given in Table 1. Hereinafter, coarse stainless-steel mesh is abbreviated as "cSS", fine stainless steel mesh as "fSS", and nickel net with a very fine mesh width as "vfNi".

Table 1. Physical data of the investigated wire meshes for electrode coating.

Mesh Type	Mesh Size (mm^2)	Wire Gauge (mm)	Surface Area (mm^2 cm^{-2})	Manufacturer
Coarse stainless steel (cSS)	1.8 × 1.8	0.32	94.8	Spörl KG, Germany
Fine stainless steel (fSS)	1.0 × 1.0	0.22	116.3	Spörl KG, Germany
Very fine nickel (vfNi)	0.106 × 0.118	0.063	627.2	Harver & Böcker, Germany

2.3. Electrode Spray Coating Process

The coating process for the cathodes is shown in Figure 3. At first, a coating solution has to be prepared, consisting of a basic polymer solution, catalyst, and additional butanone. The basic polymer binder solution was prepared by dissolving three table tennis balls (ink-free with no commercial logos) in 150 mL butanone to obtain a celluloid solution. Due to the strict standards, table tennis balls form high-quality educts at low prices and fit perfectly to the low-cost approach for a low-cost flexible binder. Then, the basic polymer solution (20 mL) was merged with additional butanone (100 mL). To obtain a homogeneous dispersion, the solution was stirred, and the catalysts were added stepwise (4 g carbon + 0.4 g additional catalyst: MnO_2 and/or MoS_2, Co_3O_4). Then, the mixture was homogenized for 15 min with a dispersing tool (Heidolph, Germany, model SilentCrusher at 13,500 rpm).

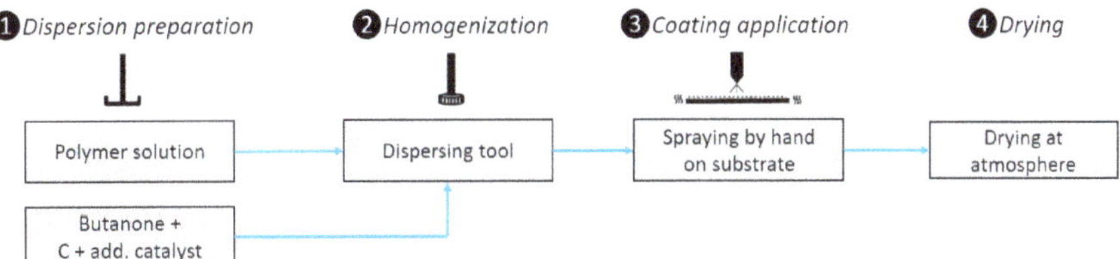

Figure 3. Flow diagram of the process steps for electrode preparation using the spray method.

A spray table with heating option and the degreased (degreaser: isopropanol) metal mesh were heated and controlled at a very stable surface temperature of 110 °C. The dispersion was manually applied with an airbrush spray gun (Harder & Steenbeck, Norderstedt, Germany, model: Evolution Solo) in cross-coat. Inert nitrogen (Linde Gases Division, Pullach, Germany) was used as carrier gas for spraying (3.5 N grade and a pressure of 2 bar). The applicated mass was controlled by weighting the cathode before, during, and after the coating process. Furthermore, the optimum catalyst loading was evaluated and finally fixed to 0.4 mg/cm^2 by preliminary tests. The coating process was completed when the desired catalyst loading was reached. No calcination step was required after coating; therefore, the coating process was timesaving and avoided further changes to the catalyst.

2.4. Measuring Procedure and Analytical Methods

A Luggin capillary is usually used to investigate the electrode potential compared with a reference electrode. In order to not only measure the potential at one geometric point, a membrane extension operated as a salt bridge between the cathode and the reference electrode. Therefore, the potentials of the electrodes were measured in an integrative way over the entire active area [41]. The membrane extension was kept moist by an applied carbon fleece in order to stabilize the ionic conductivity and the measurement (Figure 4). A reversible hydrogen electrode (RHE, Gaskatel GmbH, Kassel, Germany) served as a reference electrode, which was operated in a buffer solution at pH 7 (Libutec GmbH, Langenfeld, Germany; type: buffer solution pH 7).

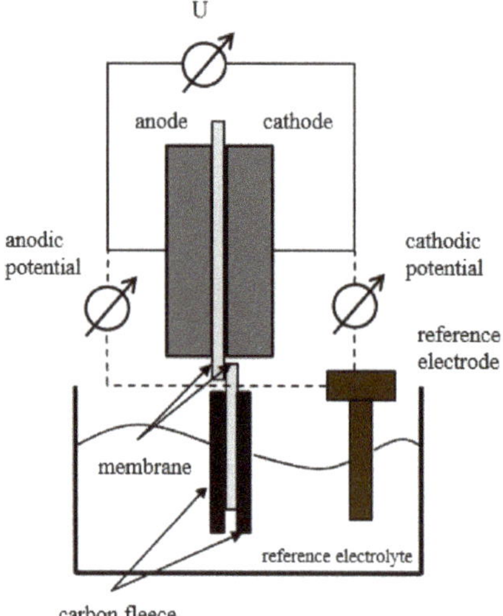

Figure 4. Sketch of the measurement setup and the salt bridge-like membrane extension for potential determination. This modified version is based on [41].

The cathodes were characterized by measuring the electrode potentials and the cell voltages as a function of the current. Furthermore, the power density characteristic curves were calculated and normalized to the geometric area. The power was adjusted between 0 mA to short circuit current in 1 mA steps. After each step, the stationary state was awaited (average duration: 5 min).

To analyze the chemical composition, an X-ray powder diffraction (XRD) analysis was conducted with a D/max-2200PC-X-ray (Rigaku Corp., The Woodlands, TX, USA) diffractometer (40 kV, 20 mA) using CuKα radiation (0.15404 nm) and a scan range between 10° to 80°, with 10°/min.

Physical properties were analyzed by the Brunauer-Emmet-Teller (BET) method with a NOVA2000e device (Micromeritics GmbH, Unterschleissheim, Germany) with a prearranged outgassing step at 200 °C at a vacuum pressure of 6 mmHg of the samples. Selected cathodes were examined with a scanning electron microscope (SEM) at the Institute of Mechanical Process Engineering at Clausthal University of Technology (Carl Zeiss Microscopy GmbH, Jena, Germany, model: DSM 982 Gemini.M).

3. Results and Discussion

3.1. Characterization of Catalysts and Supports

MnO_2 and MoS_2 have already been evaluated using XRD and BET; reference is made to [9]. It should be noted that gamma MnO_2 was intentionally used, which is significantly cheaper than other modifications. Crystalline phases of the investigated TiO_2 supports, and its combination with Co_3O_4, are shown in Figure 5. The whiskers type of TiO_2 was abbreviated as TiO_2-W, whereas TiO_2-A is the abbreviation for the anatase modification. The XRD pattern of the supports showed peaks at 2θ = 25.2°, 37.8°, 48.0°, and 55.0°, confirming both modifications were anatase phase TiO_2. The analysis of the catalyst-loaded TiO_2 supports revealed a successful loading, as the peaks of TiO_2 were still present, next to the peaks of Co_3O_4 at 2θ = 36.9° and 65.2°.

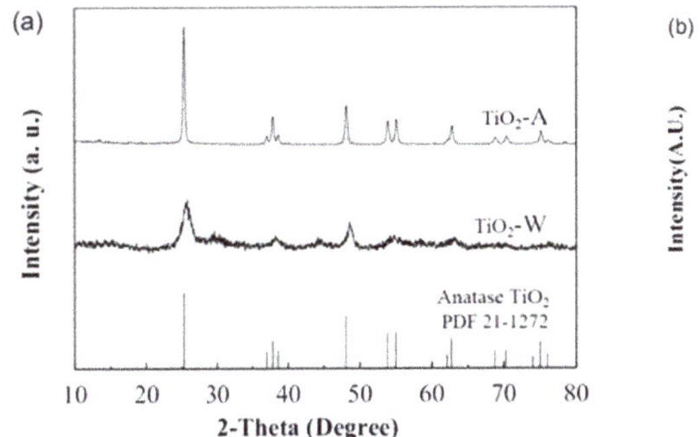

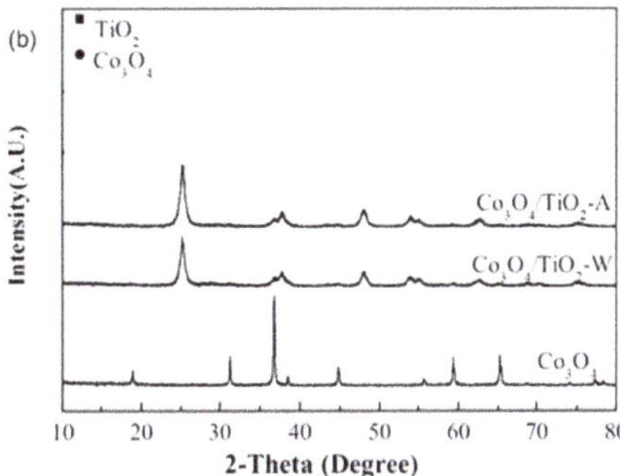

Figure 5. X-ray powder diffraction (XRD) pattern of TiO_2 supports (**a**) and in combination with Co_3O_4 catalyst (**b**).

The BET surface areas of the TiO_2-supported and prepared Co_3O_4/TiO_2 catalysts are presented in Table 2. The comparison between the investigated TiO_2 modifications showed that the TiO_2-W had a higher specific surface area (85.9 to 70 m^2/g), and larger pore volume and size. After catalyst loading, the surface area decreased, but the loaded TiO_2-W support still had a greater surface area.

Table 2. Catalyst content and physical properties of TiO_2 carrier and its combination with Co_3O_4. TiO_2-A: anatase modification of TiO_2, TiO_2-W: whiskers type of TiO_2.

Sample	S_{BET} ($m^2\ g^{-1}$)	V_P ($cm^3\ g^{-1}$)	*d* (nm)	*Surface Content* (weight %)
TiO_2-A	70.0	0.241	13.5	-
TiO_2-W	85.9	0.400	18.3	-
Co_3O_4/TiO_2-A	60.9	0.231	14.1	13.0
Co_3O_4/TiO_2-W	77.8	0.323	15.9	4.2

Figure 6 illustrates the Co 2p3/2 spectra of the prepared catalysts. All patterns revealed two peaks at 786.2 eV and 779.4 eV, whereas the small peak can be attributed to Co^{2+} species, and the bigger peak at 779.4 eV to Co^{3+}. Using this reasoning, Co_3O_4 was the predominant phase for all samples.

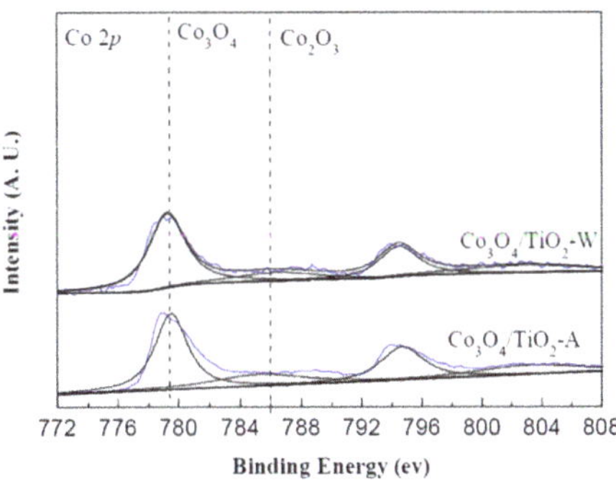

Figure 6. Co 2p3/2 spectra for the supported catalysts. The binding energy for Co_3O_4 (779.4 eV) and Co_2O_3 (786.2 eV) is shown.

3.2. Performance of Coated Electrodes

3.2.1. Comparison of MnO_2, MoS_2, and Co_3O_4-Based Cathodes

The presented spraying method was applied to a variety of catalysts (MnO_2, MoS_2, Co_3O_4, and Pt) and different metal meshes. A summary of the fabricated and tested combinations is given in Table 3.

Table 3. Overview of the produced electrodes by the novel spray method on different meshes.

Added Catalyst	Coarse Stainless Steel Mesh	Fine Stainless Steel Mesh	Very Fine Ni Mesh
Pt	x		
MnO_2	x	x	x
MoS_2	x	x	x
MnO_2 + MoS_2 with 1:1	x	x	x
MnO_2 + MoS_2 with 1:2	x	x	x
MnO_2 + MoS_2 with 2:1	x	x	x
Co_3O_4 at TiO_2-A	x		
Co_3O_4 at TiO_2-W	x		

In order to achieve a concise comparison of the investigated cathodes, all electrode configurations were characterized as described. Figure 7 shows an example of a power density curve, including the anode and cathode potential curve, as well as the current–voltage curve.

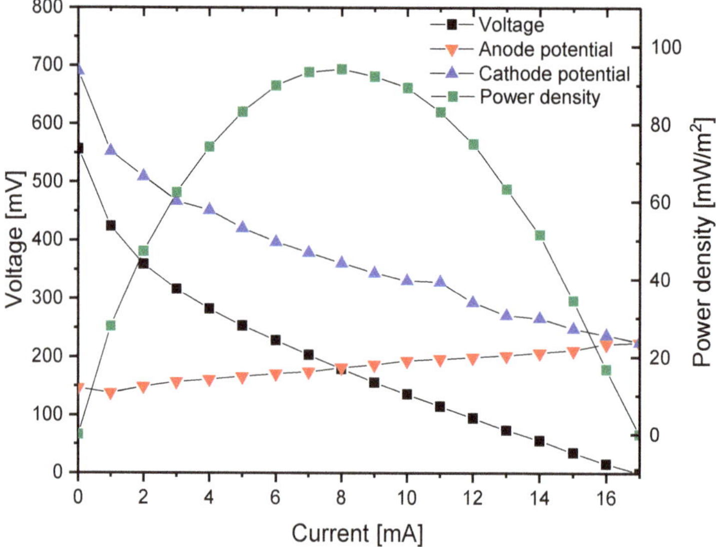

Figure 7. Characteristic curves of an microbial fuel cell (MFC) with a spray-coated cathode based on MnO_2-MoS_2 (1:1) on fSS wire mesh. This example also underlines the cathodic bottleneck of the system, determined by the stronger decreasing cathodic potential in comparison to the anodic potential.

The maximum performance of each electrode is given in Figure 8. For the sake of clarity, electrodes with different meshes, but identical catalysts, are presented in clusters. A cathode based on the reference catalyst Pt (fuel cell grade Pt at cSS substrate) was produced and used for comparison in Figure 8 (left).

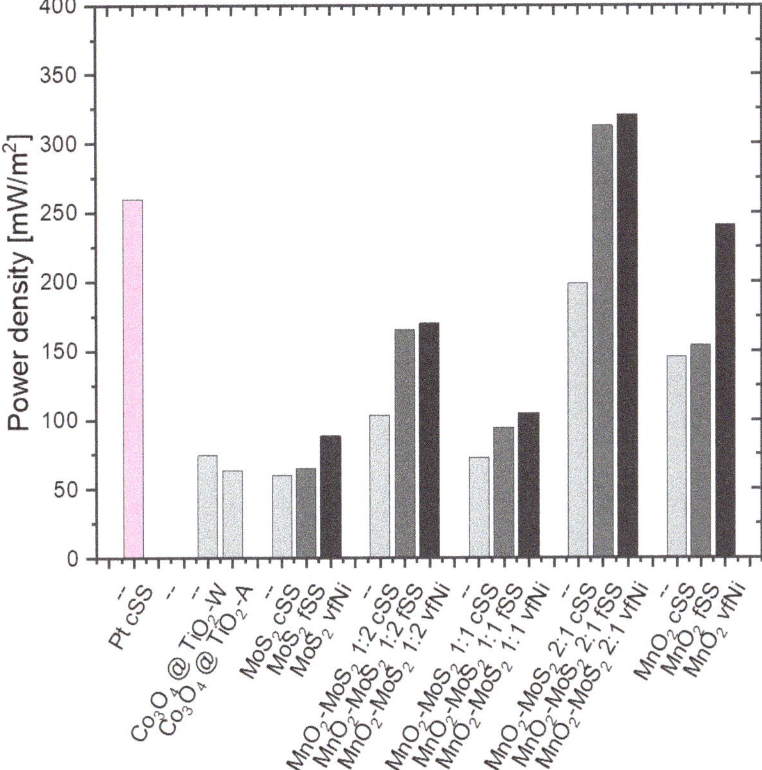

Figure 8. Comparison of maximum power densities of different cathodes. All electrodes were manufactured with a catalyst load of 0.4 mg/cm^2.

The comparison clearly indicates that all catalysts benefit from an increasing surface area of the metal meshes. The performance increases from coarse (cSS) to fine stainless-steel mesh (fSS) and is outperformed by the cathodes on very fine nickel mesh (vfNi). However, the 5.4-times increase of surface area of vfNi mesh cathode compared to fSS mesh cathodes only led to a slight increase in MFC performance for different catalyst systems, including the best one (MnO_2-MoS_2 2:1). For the MnO_2 catalyst, a significant increase of approximately 80 mW/m^2 (30%) was found, but still at a lower level than the best catalyst system. These results indicate an existing limit for the impact of electrode surface area on the improvement of MFC performance.

Electrodes with CO_3O_4 (on TiO_2 support) and MoS_2 showed the lowest power density (only 60–88 mW/m^2). Both investigated Co_3O_4 on TiO_2 systems (in combination with carbon and polymer) showed similar results, but the Co_3O_4/TiO_2-W cathode was slightly better (74.4 mW/m^2) in comparison to Co_3O_4/TiO_2-A (63.6 mW/m^2), probably caused by the increased surface area.

The increasing surface area caused by the finer mesh width also had a positive effect on MoS_2-based cathodes, but the performance was still low (60.2, 65.1, and 88.5 mW/m^2). The oxygen reduction capability of MnO_2 was considerably higher. The increase from cSS to fSS enhanced the maximum power density from 145.4 to 153.9 mW/m^2. A further improvement was given by the usage of the vfNi mesh, showing power densities up to 240.9 mW/m^2. Sulphur compounds in wastewater are known as catalyst poisons, and it was recently shown that a catalyst combination with MoS_2 increases the long-term stability of MFC cathode catalysts [9]. Therefore, different mixing ratios were investigated.

The (weight %) mixing ratio of 1:1 (MnO_2-MoS_2) led to a performance deterioration compared to the pure MnO_2 system, and the electrodes showed similar performances to the MoS_2 cathodes on all substrates. The mixing ratios of 1:2 and 2:1 performed differently, but they showed an improving effect. In particular, the electrodes with a mixing ratio of 2:1 showed power densities of 198.3 mW/m^2 (MnO_2-MoS_2 2:1, cSS), 312.3 mW/m^2 (MnO_2-MoS_2 2:1, fSS), and 320.6 mW/m^2 (MnO_2-MoS_2 2:1, vfNi). An even better performance was reached with the fSS and vfNi meshes than for the Pt-coated cathode. Potential/current curves of selected cathodes are shown in Figure 9. The diagram clearly indicates the most stable cathode potential curve for MnO_2/MoS_2 (2:1). Furthermore, the cathode potential stabilized in terms of current stability from the cSS mesh (grey curve) over the fSS mesh (blue curve) to the vfNi mesh (green curve), with simultaneously increasing cathode surface area. All cathodes showed a potential drop, especially at higher currents. This was probably caused by mass transport limitations.

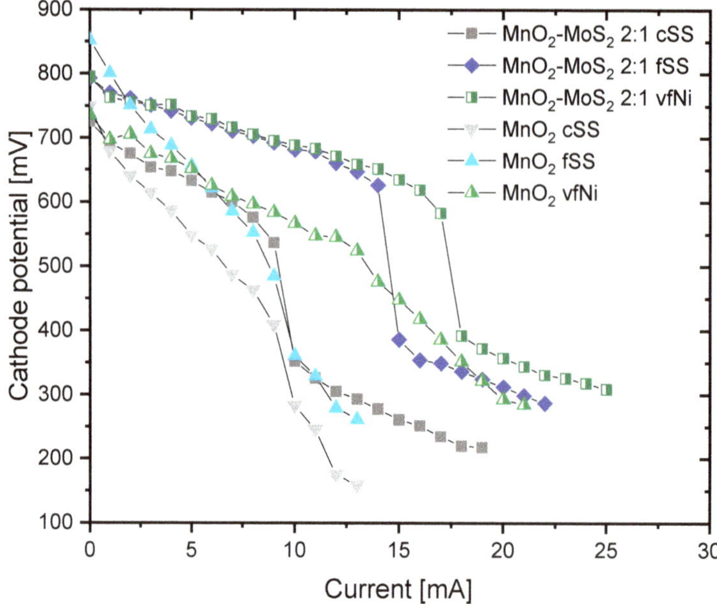

Figure 9. Cathode potentials of selected spray-coated cathodes on different substrates.

Due to the high costs for Co_3O_4, carrier-supported catalysts (Co_3O_4 at TiO_2) were evaluated. However, these showed low ORR activity and the achieved currents were below published levels. Obviously, pure Co_3O_4 is expected to show significantly better performance, but this is not suitable for wastewater treatment by MFC due to the low cost-efficiency. In conclusion, the MnO_2-based cathodes showed significantly better performances. Especially a combination with MoS_2 could surpass the performances related to scalable systems described in the literature to date.

3.2.2. Comparison to Paintbrush-Manufactured Cathodes

Novel spray coated cathodes were compared to cathodes previously produced by the paintbrush method [9] by SEM. All cathodes had a catalyst ratio of 2:1 (MnO_2/MoS_2). According to Figure 10, the paintbrush process produced a tight fit between the interstitial spaces, whereby some were closed, whereas other interstitial spaces remained empty. The coating was not uniform and showed a tendency for detaching. The application consisted mainly of elongated particles attached to each other in a porous structure, which were fixed in position by form closure.

Figure 10. Scanning electron microscope (SEM) image of a cathode produced by paintbrush method with a catalyst ratio of 2:1 (MnO_2/MoS_2) in different scales.

In contrast, the cathodes produced by the spray coating method showed a more regular coating on the surface of the wire mesh. The finer the wire mesh, the more uniform the distribution. The SEM image of coated vfNi substrate is shown as an example in Figure 11.

Figure 11. SEM image of a spray coated cathode with a catalyst ratio of 2:1 (MnO_2/MoS_2) in different scales.

Thus, it was shown that the presented spray method enabled uniform and reproducible catalytic coatings. The developed spray coating method seems to be a promising process for the production of high-performance electrodes and also for upscaling. The method also enabled a suitable comparison between different catalysts through homogenous coatings. No additional calcination steps were necessary, which could modify the catalyst structure.

3.2.3. Long-Term Performance of Selected Electrodes

The long-term performance of the best performing spray-coated cathode (MnO_2-MoS_2 2:1 vfNi) in comparison to a paintbrush-coated cathode (according to [9]) is given in Figure 12. It was noted that the paintbrush method was applied on a cSS net, as the paintbrush method would cause clogging on finer mesh width. Although a drop in power density was noticeable at the beginning of the measurement, a relatively constant power density of 150 mW/m^2 was reached from the fourth day onwards. From day 6, the power density rose, reaching 200–240 mW/m^2 from day 10 onwards. The long-term performance of paintbrush-coated electrode was about 60 to 100 mW/m^2. The comparison of the long-term performance between both electrodes by different coating methods revealed a performance improvement by factor 2 to 3, or in terms of power density, by 100 to 180 mW/m^2 for the spray-coated cathode. The comparison between the improvement of long-term power density and the improvement of maximum power density showed that the improvement in long-term performance was higher than that in maximum performance. Therefore, the spray-coating method was identified as enabling the production of high-performing and long-term stable cathodes for MFC application.

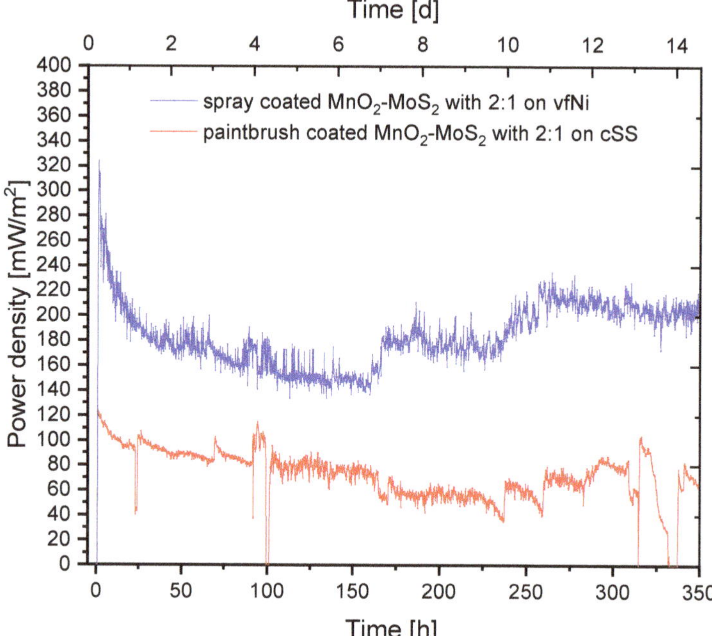

Figure 12. Long-term performance of the spray-coated MnO_2-MoS_2 (2:1) vfNi cathode.

4. Conclusions

To scale-up MFCs to the dimensions of technical-scale wastewater treatment plants, it is necessary to manufacture scalable electrodes. Therefore, the composition of a universal sprayable suspension was developed within this study, which was used for a simplified airbrush spray-coating method. The performance was evaluated regarding different catalysts (Co_3O_4, MnO_2, MoS_2, and selected combinations) on different carrier meshes and materials. The spray-coating method facilitated a homogeneous coating and stood out with an increased cathode performance. This could be enhanced further by suitable substrates with a finer mesh width and a higher surface area. The most promising cathode catalyst composition of the tested systems was the combination of MnO_2 and MoS_2 mixture (2:1) with a maximum and long-term power density of 320 and 200–240 mW/m^2, respectively. The coated electrodes also demonstrated long-term stability. This contribution confirmed the effect of the quality of the catalyst coating of cathodes for wastewater treatment by microbial fuel cells (MFC). A promising and more rapid manufacturing method for better catalyst comparisons and large-scale applications was identified. This process could also be the basis for an automated coating. Moreover, this work demonstrated that not only the catalytic components and their composition influenced electrocatalytic properties. The method of preparation and the carrier mesh also exerted a large effect on the electrode performance.

Author Contributions: Conceptualization, methodology, validation, data curation, visualization, writing—original draft preparation, T.M. and D.H.; investigation, T.M., D.H., and B.J.; writing—review and editing, all authors; supervision, project administration, funding acquisition, M.S. and U.K. All authors have read and agreed to the published version of the manuscript.

Funding: This research was funded by the Federal Ministry of Education and Research (Bundesministerium für Bildung und Forschung), BMBF, Germany, grant number WTER0219813.

Conflicts of Interest: The authors declare no conflict of interest.

References

1. Muddemann, T.; Haupt, D.; Sievers, M.; Kunz, U. Electrochemical Reactors for Wastewater Treatment. *ChemBioEng Rev.* **2019**, *6*, 142–156. [CrossRef]
2. United Nations. *The Millennium Development Goals Report*; United Nations: New York, NY, USA, 2006.
3. UN Water and FAO. *Coping with Water Scarcity—Challenge of the Twenty-First Century*; UN Water and FAO: Rome, Italy, 2007.
4. Logan, B.E. *Microbial Fuel Cells*; John Wiley & Sons, Inc.: Hoboken, NJ, USA, 2008; ISBN 978-0-470-23948-3.
5. Hernández-Fernández, F.J.; Pérez de los Ríos, A.; Salar-García, M.J.; Ortiz-Martínez, V.M.; Lozano-Blanco, L.J.; Godínez, C.; Tomás-Alonso, F.; Quesada-Medina, J. Recent progress and perspectives in microbial fuel cells for bioenergy generation and wastewater treatment. *Fuel Process. Technol.* **2015**, *138*, 284–297. [CrossRef]
6. Zhang, X.; He, W.; Ren, L.; Stager, J.; Evans, P.J.; Logan, B.E. COD removal characteristics in air-cathode microbial fuel cells. *Bioresour. Technol.* **2015**, *176*, 23–31. [CrossRef] [PubMed]
7. Malvankar, N.S.; Lovley, D.R. Microbial nanowires: A new paradigm for biological electron transfer and bioelectronics. *ChemSusChem* **2012**, *5*, 1039–1046. [CrossRef] [PubMed]
8. Yuan, H.; Hou, Y.; Abu-Reesh, I.M.; Chen, J.; He, Z. Oxygen reduction reaction catalysts used in microbial fuel cells for energy-efficient wastewater treatment: A review. *Mater. Horiz.* **2016**, *3*, 382–401. [CrossRef]
9. Jiang, B.; Muddemann, T.; Kunz, U.; Silva e Silva, L.G.; Bormann, H.; Niedermeiser, M.; Haupt, D.; Schläfer, O.; Sievers, M. Graphite/MnO_2 and MoS_2 Composites Used as Catalysts in the Oxygen Reduction Cathode of Microbial Fuel Cells. *J. Electrochem. Soc.* **2017**, *164*, E519–E524. [CrossRef]
10. Rahimnejad, M.; Adhami, A.; Darvari, S.; Zirepour, A.; Oh, S.-E. Microbial fuel cell as new technology for bioelectricity generation: A review. *Alex. Eng. J.* **2015**, *54*, 745–756. [CrossRef]
11. Rismani-Yazdi, H.; Carver, S.M.; Christy, A.D.; Tuovinen, O.H. Cathodic limitations in microbial fuel cells: An overview. *J. Power Sources* **2008**, *180*, 683–694. [CrossRef]
12. Santoro, C.; Arbizzani, C.; Erable, B.; Ieropoulos, I. Microbial fuel cells: From fundamentals to applications. A review. *J. Power Sources* **2017**, *356*, 225–244. [CrossRef]
13. Muddemann, T.; Haupt, D.R.; Gomes Silva e Silva, L.; Jiang, B.; Kunz, U.; Bormann, H.; Niedermeiser, M.; Schlaefer, O.; Sievers, M. Integration of Upscaled Microbial Fuel Cells in Real Municipal Sewage Plants. *ECS Trans.* **2017**, *77*, 1053–1077. [CrossRef]
14. Jiang, B.; Muddemann, T.; Kunz, U.; Bormann, H.; Niedermeiser, M.; Haupt, D.; Schläfer, O.; Sievers, M. Evaluation of Microbial Fuel Cells with Graphite Plus MnO_2 and MoS_2 Paints as Oxygen Reduction Cathode Catalyst. *J. Electrochem. Soc.* **2016**, *164*, H3083–H3090. [CrossRef]

15. Kumar, R.; Singh, L.; Zularisam, A.W. Enhanced oxygen reduction reaction in air-cathode microbial fuel cells using flower-like Co_3O_4 as an efficient cathode catalyst. *Int. J. Hydrog. Energy* **2017**, *42*, 19287–19295. [CrossRef]
16. Li, X.; Hu, B.; Suib, S.; Lei, Y.; Li, B. Manganese dioxide as a new cathode catalyst in microbial fuel cells. *J. Power Sources* **2010**, *195*, 2586–2591. [CrossRef]
17. Li, X.; Hu, B.; Suib, S.; Lei, Y.; Li, B. Electricity generation in continuous flow microbial fuel cells (MFCs) with manganese dioxide (MnO_2) cathodes. *Biochem. Eng. J.* **2011**, *54*, 10–15. [CrossRef]
18. Nandy, A.; Sharma, M.; Venkatesan, S.; Taylor, N.; Gieg, L.; Thangadurai, V. Comparative Evaluation of Coated and Non-Coated Carbon Electrodes in a Microbial Fuel Cell for Treatment of Municipal Sludge. *Energies* **2019**, *12*, 1034. [CrossRef]
19. Zhang, Y.; Liu, L.; van der Bruggen, B.; Yang, F. Nanocarbon based composite electrodes and their application in microbial fuel cells. *J. Mater. Chem. A* **2017**, *5*, 12673–12698. [CrossRef]
20. Hao, L.; Yu, J.; Xu, X.; Yang, L.; Xing, Z.; Dai, Y.; Sun, Y.; Zou, J. Nitrogen-doped MoS_2/carbon as highly oxygen-permeable and stable catalysts for oxygen reduction reaction in microbial fuel cells. *J. Power Sources* **2017**, *339*, 68–79. [CrossRef]
21. Kodali, M.; Santoro, C.; Serov, A.; Kabir, S.; Artyushkova, K.; Matanovic, I.; Atanassov, P. Air Breathing Cathodes for Microbial Fuel Cell using Mn-, Fe-, Co- and Ni-containing Platinum Group Metal-free Catalysts. *Electrochim. Acta* **2017**, *231*, 115–124. [CrossRef]
22. Tsai, H.-Y.; Hsu, W.-H.; Liao, Y.-J. Effect of Electrode Coating with Graphene Suspension on Power Generation of Microbial Fuel Cells. *Coatings* **2018**, *8*, 243. [CrossRef]
23. Satar, I.; Daud, W.R.W.; Kim, B.H.; Somalu, M.R.; Ghasemi, M.; Bakar, M.H.A.; Jafary, T.; Timmiati, S.N. Performance of titanium–nickel (Ti/Ni) and graphite felt-nickel (GF/Ni) electrodeposited by Ni as alternative cathodes for microbial fuel cells. *J. Taiwan Inst. Chem. Eng.* **2018**, *89*, 67–76. [CrossRef]
24. Hideki, M. Manufacturing Equipment and Manufacturing Method of Membrane Electrode Assembly. US20100051181A1, 4 March 2010.
25. Kwon, N.H.; Hwang, I.C.; Ahn, B.K.; Lim, T.W. Method and Apparatus for Preparing Catalyst Slurry for Fuel Cells. US20100086450A1, 8 April 2010.
26. Tochigi, T.T.; Tochigi, S.T.; Tochigi, K.Y. Verfahren zur Herstellung einer Elektrodenschicht für eine Brennstoffzelle. DE1020006046373A1, 5 April 2007.
27. Bulan, A. Sauerstoffverzehrelektrode und Verfahren zu ihrer Herstellung. DE102010024053A1, 22 December 2011.
28. Kintrup, J.; Bulan, A.; Hammarberg, E.; Sepeur, S.; Frenzer, G.; Gross, F. Beschichtungsmischung Herstellung Covestro. US2017067172A1, 9 March 2017.
29. Yang, W.; Li, J.; Lan, L.; Fu, Q.; Zhang, L.; Zhu, X.; Liao, Q. Poison tolerance of non-precious catalyst towards oxygen reduction reaction. *Int. J. Hydrog. Energy* **2018**, *43*, 8474–8479. [CrossRef]
30. Zhang, L.; Liu, C.; Zhuang, L.; Li, W.; Zhou, S.; Zhang, J. Manganese dioxide as an alternative cathodic catalyst to platinum in microbial fuel cells. *Biosens. Bioelectron.* **2009**, *24*, 2825–2829. [CrossRef] [PubMed]
31. Roche, I.; Katuri, K.; Scott, K. A microbial fuel cell using manganese oxide oxygen reduction catalysts. *J. Appl. Electrochem.* **2010**, *40*, 13–21. [CrossRef]
32. Zhou, X.; Xu, Y.; Mei, X.; Du, N.; Jv, R.; Hu, Z.; Chen, S. Polyaniline/β-MnO_2 nanocomposites as cathode electrocatalyst for oxygen reduction reaction in microbial fuel cells. *Chemosphere* **2018**, *198*, 482–491. [CrossRef]
33. Majidi, M.R.; Shahbazi Farahani, F.; Hosseini, M.; Ahadzadeh, I. Low-cost nanowired α-MnO_2/C as an ORR catalyst in air-cathode microbial fuel cell. *Bioelectrochemistry* **2018**, *125*, 38–45. [CrossRef]
34. Shahbazi Farahani, F.; Mecheri, B.; Reza Majidi, M.; Costa de Oliveira, M.A.; D'Epifanio, A.; Zurlo, F.; Placidi, E.; Arciprete, F.; Licoccia, S. MnO_x-based electrocatalysts for enhanced oxygen reduction in microbial fuel cell air cathodes. *J. Power Sources* **2018**, *390*, 45–53. [CrossRef]
35. Woon, C.W.; Ong, H.R.; Chong, K.F.; Chan, K.M.; Khan, M.M.R. MnO_2/CNT as ORR Electrocatalyst in Air-Cathode Microbial Fuel Cells. *Procedia Chem.* **2015**, *16*, 640–647. [CrossRef]
36. Kumar, R.; Singh, L.; Zularisam, A.W.; Hai, F.I. Potential of porous Co_3O_4 nanorods as cathode catalyst for oxygen reduction reaction in microbial fuel cells. *Bioresour. Technol.* **2016**, *220*, 537–542. [CrossRef]
37. Jiang, B. Optimization of Cathode Performance of Microbial Fuel Cells for Wastewater Treatment and Electrochemical Power Evaluation. Ph.D. Thesis, Clausthal University of Technology, Clausthal-Zellerfeld, Germany, 2018.

38. Rossi, R.; Jones, D.; Myung, J.; Zikmund, E.; Yang, W.; Gallego, Y.A.; Pant, D.; Evans, P.J.; Page, M.A.; Cropek, D.M.; et al. Evaluating a multi-panel air cathode through electrochemical and biotic tests. *Water Res.* **2018**, *148*, 51–59. [CrossRef]
39. Brown, R.K.; Harnisch, F.; Wirth, S.; Wahlandt, H.; Dockhorn, T.; Dichtl, N.; Schröder, U. Evaluating the effects of scaling up on the performance of bioelectrochemical systems using a technical scale microbial electrolysis cell. *Bioresour. Technol.* **2014**, *163*, 206–213. [CrossRef]
40. Chen, S.; Zhu, Y.; Li, W.; Liu, W.; Li, L.; Yang, Z.; Liu, C.; Yao, W.; Lu, X.; Feng, X. Synthesis, Features, and Applications of Mesoporous Titania with TiO_2(B). *Chin. J. Catal.* **2010**, *31*, 605–614. [CrossRef]
41. He, W.; van Nguyen, T. Edge Effects on Reference Electrode Measurements in PEM Fuel Cells. *J. Electrochem. Soc.* **2004**, *151*, A185. [CrossRef]

Article

Preparation and Characterization of Porous $Ti/SnO_2–Sb_2O_3/PbO_2$ Electrodes for the Removal of Chloride Ions in Water

Kangdong Xu [1], Jianghua Peng [2], Pan Chen [1], Wankai Gu [1], Yunbai Luo [1] and Ping Yu [1,*]

1 College of Chemistry and Molecular Sciences, Wuhan University, Wuhan 430072, China; xukangdong0520@163.com (K.X.); 2016202030162@whu.edu.cn (P.C.); wiki_gu@163.com (W.G.); ybai@whu.edu.cn (Y.L.)
2 MOE Key Laboratory of Thermo-Fluid Science and Engineering, Xi'an Jiaotong University, Xi'an 710049, China; pengke112@126.com
* Correspondence: yuping@whu.edu.cn; Tel.: +86-027-6875-2511

Received: 14 September 2019; Accepted: 14 October 2019; Published: 18 October 2019

Abstract: Porous $Ti/SnO_2–Sb_2O_3/PbO_2$ electrodes for electrocatalytic oxidation of chloride ions were studied by exploring the effects of different operating conditions, including pore size, initial concentration, current density, initial pH, electrode plate spacing, and the number of cycles. In addition, a physicochemical characterization and an electrochemical characterization of the porous $Ti/SnO_2–Sb_2O_3/PbO_2$ electrodes were performed. The results showed that $Ti/SnO_2–Sb_2O_3/PbO_2$ electrodes with 150 μm pore size had the best removal effect on chloride ions with removal ratios amounting up to 98.5% when the initial concentration was 10 g L^{-1}, the current density 125 mA cm^{-2}, the initial pH = 9, and the electrode plate spacing 0.5 cm. The results, moreover, showed that the oxygen evolution potential of 150 μm porous $Ti/SnO_2-Sb_2O_3/PbO_2$ electrodes was highest, which minimized side reactions involving oxygen formation and which increased the removal effect of chloride ions.

Keywords: electrocatalytic oxidation; chloride ions removal ratio; the porous electrode; influencing factors

---

1. Introduction

In recent years, China has raised the level of environmental protection, which requires industrial wastewaters not to be discharged from various enterprises. Therefore, in order to ensure the normal operation of its own production, each enterprise must realize the recycling of water. At present, the commonly used method is to properly dispose of the drainage and then replenish it into the industrial water system. However, various ions are continuously enriched in the water when the water is reused, which causes various adverse effects on the operation of equipment. The amount of scale cations (such as calcium ions and magnesium ions, etc.) in water can be reduced by changing the pH of the water body and by inducing flocculation sedimentation [1], but there is no effective method of reducing anions (such as chloride ions) which have corrosive effects on equipment in water [2]. Therefore, there is a need to find a way to quickly and easily reduce chloride ions in water.

At present, the methods for removing chloride ions include biological [3,4], reverse osmosis [5,6], distillation, multi-effect evaporation [7], electrodialysis [8–10], and ions exchange methods. The biological method involves high operating costs, while the removal effect does not work well. The reverse osmosis method is burdened by high consumption of acid and alkali as well as energy consumption. The distillation method and the multi-effect evaporation methods involve high energy consumption. The electrodialysis method involves high operation voltages, and membranes are easily contaminated. The ion exchange method is expensive and tends to cause secondary pollution during the elution process. The electrocatalytic oxidation technology is a new method for removing chloride ions, which has the advantages of high efficiency in removing chloride ions, simple process, simple operation, and low operating cost, and there are few related studies.

In this study, porous $Ti/SnO_2–Sb_2O_3/PbO_2$ electrodes were studied and it was found that these have a significant effect on electrocatalytic oxidation of removing chloride ions. Self-made porous $Ti/SnO_2–Sb_2O_3/PbO_2$ electrodes with different pore sizes were used as anodes to construct an electrolyzer. The high potential of the anode and unique catalytic oxidation characteristics of the surface coating were used for removing chloride ions. Three different kinds of porous $Ti/SnO_2–Sb_2O_3/PbO_2$ electrodes were prepared and characterized systematically with regard to morphology, crystal structure, and various electrochemical performances. NaCl solutions were used for simulating chlorine-containing wastewaters for studying the effects of initial concentration, pore size, current density, initial pH, electrode plate spacing, and the number of cycles. The results showed that the porous $Ti/SnO_2–Sb_2O_3/PbO_2$ electrodes were efficiently reducing the number of chloride ions in water, thus providing a practical method of removing chloride ions from water.

2. Experimental

2.1. Materials and Reagents

The substrates used in this study were commercial samples of porous Ti (20 mm × 10 mm × 1 mm), which had average pore sizes of 50 μm, 100 μm, and 150 μm. In regard to removing chloride ions, porous Ti substrates with larger pore sizes do not meet the requirements for removing chloride ions in terms of hardness, our studies were restricted to the three pore sizes mentioned above. In this paper, all chemicals were analytical grade without any other impurities. $SnCl_4·5H_2O$, $SbCl_3$, HCl, $Pb(NO_3)_2$, $Cu(NO_3)_2·3H_2O$, NaF, HNO_3, NaOH, $H_2C_2O_4$, NaCl, and other chemicals were obtained from Shanghai Wo Kai Biotechnology Co., Ltd. (Shanghai, China). All the solutions used in these experiments were prepared with deionized water. For the simulation of chloride ion contaminated waste waters, NaCl solutions were used.

2.2. Electrode Preparation

First, at a temperature of 70 °C, porous Ti substrates (50 μm, 100 μm, and 150 μm) were heated in sodium hydroxide (20% m%) for 1 h to remove all traces of oil on the surface and were then washed in deionized water. Thereafter, the porous Ti substrates were etched in oxalic acid (15% m%) at a temperature of 85 °C for 1 h to obtain a uniformly rough surface. Finally, the samples were washed in deionized water and stored in deionized water [11,12].

Second, the coating solution, consisting of 1.20 g $SnCl_4·5H_2O$ and 0.20 g $SbCl_3$ was dissolved in 40 mL ethanol, and 1 mL of concentrated hydrochloric acid was added. The treated porous Ti substrates were immersed in the coating solution for 5 min, and then dried at about 120 °C for 15 min, and thereafter calcined at 500 °C for 20 min in a muffle furnace. All above processes were repeated ten times and the electrodes were annealed at 500 °C for 60 min in the last process [13,14]. Finally, the porous $Ti/SnO_2–Sb_2O_3$ electrode was prepared.

Third, the deposition solution consisted of 40 g $Pb(NO_3)_2$, 15 g $Cu(NO_3)_2·3H_2O$, 0.5 g NaF, and 0.1 M HNO_3. The porous $Ti/SnO_2–Sb_2O_3$ electrodes were used as anodes and a pure titanium plate as a cathode. The current density was 5 mA cm^{-2} and the PbO_2 was deposited on the porous $Ti/SnO_2–Sb_2O_3$ electrode at 65 °C for 0.5 h under stirring conditions. Finally, the prepared porous $Ti/SnO_2–Sb_2O_3/PbO_2$ electrodes were washed in deionized water [15–17].

2.3. Electrode Characterization

The surface morphologies of porous $Ti/SnO_2–Sb_2O_3/PbO_2$ electrodes were characterized by Field Emission Scanning Electron Microscopy (Zeiss SIGMA, Carl Zeiss Corporation, Jena, Germany). The composition and the chemical state of the porous Ti/SnO_2–Sb_2O_3/PbO_2 electrode was determined by an X-ray photoelectron spectrometer (ESCALAB250Xi, Thermo Fisher Scientific, Waltham, MA, USA). The crystal structure of porous Ti/SnO_2–Sb_2O_3/PbO_2 electrodes were determined by an X-ray Diffractometer (XRD–6100, Shimadzu Corporation, Kyoto, Japan) with Cu–Kα (λ = 0.154 nm) incident radiation at a scanning rate of 2 min^{-1} in 2θ mode from 20 to 90.

The electrochemical characterization including linear sweep voltammetry (LSV) curves and cyclic voltammetry of porous Ti/SnO_2–Sb_2O_3/PbO_2 electrodes were carried out using a CHI760E electrochemical workstation (CH Instruments, Chenhua Co., Shanghai, China) with a conventional three-electrode cell. The working electrode was the PbO_2 electrode; the reference electrode was a saturated calomel electrode (SCE) and a platinum electrode served as a counter electrode.

2.4. Electrocatalytic Oxidation of Chloride Ions

The porous Ti/SnO_2–Sb_2O_3/PbO_2 electrodes were used as anodes and a Ti electrode as the cathode. The produced chlorine was absorbed in an absorption tank. After the reaction, $AgNO_3$ solution was used as a titrant and K_2CrO_4 solution as a color developer. In order to obtain optimal removal conditions, the effects of the following factors were considered for electrocatalytic oxidation of chloride ions, including initial concentration (from 5 g L^{-1} to 25 g L^{-1}), pore size (from 50 μm to 150 μm), current densities (from 50 mA cm^{-2} to 125 mA cm^{-2}), initial pH (from 3 to 11), and electrode plate spacing (from 0.5 cm to 1 cm). The removal ratio of chloride ions was calculated as follows:

$$\text{Removal ratios of chloride ions} = \frac{B_1 - B_2}{B_1} \times 100\%$$

B_1 is the concentration of the original chloride ions in the NaCl solution, and B_2 the concentration of the remaining chloride ions in the NaCl solution after the reaction.

3. Results and Discussion

3.1. Physicochemical Characterization

3.1.1. Scanning Electron Microscopy (SEM) Characterization

In Figure 1, the scanning electron microscopy (SEM) images of the three different pore sizes (50 μm, 100 μm, and 150 μm) are shown. As shown in Figure 1a,d,g, the surfaces of the porous Ti substrates are very rough. Figure 1b,e,h show that the porous Ti/SnO_2–Sb_2O_3 electrodes still have a lot of irregular pores, which indicates that the porous Ti/SnO_2–Sb_2O_3 electrodes have large specific surface areas. By zooming 500 and 4000 times, it is seen that the SnO_2–Sb_2O_3 intermediate layers are uniformly distributed and crack-free, which is beneficial for prolonging the service life of the electrode. Figure 1c,f,i show that the porous Ti/SnO_2–Sb_2O_3/PbO_2 electrodes still have many irregular pores. Although some of the pores became during electrodeposition, large specific surface areas were still retained compared to the planar electrodes. By zooming 500 and 4000 times, it is revealed that the PbO_2 grains are also very compact and evenly distributed. In short, the porous Ti/SnO_2–Sb_2O_3/PbO_2 electrodes have large surface areas, which can provide many active sites for electrochemical oxidation [18].

Figure 1. Scanning electron microscopy (SEM) images: (**a**) 50 μm porous Ti substrate; (**b**) 50 μm porous Ti/SnO_2–Sb_2O_3 electrode; (**c**) 50 μm porous Ti/SnO_2–Sb_2O_3/PbO_2 electrode; (**d**) 100 μm porous Ti substrate; (**e**) 100 μm porous Ti/SnO_2–Sb_2O_3 electrode; (**f**) 100 μm porous Ti/SnO_2–Sb_2O_3/PbO_2 electrode; (**g**) 150 μm porous Ti substrate; (**h**) 150 μm porous Ti/SnO_2–Sb_2O_3 electrode; (**i**) 150 μm porous Ti/SnO_2–Sb_2O_3/PbO_2 electrode.

3.1.2. X-ray Photoelectron Spectrometer (XPS) Characterization

In order to investigate the chemical state of each element in the porous Ti/SnO_2–Sb_2O_3/PbO_2 electrodes, the electrodes were analyzed by XPS. Since the XPS spectra of the three different-pore-size electrodes are the same, only the 150 μm porous Ti/SnO_2–Sb_2O_3/PbO_2 electrode is analyzed here. Figure 2a is the full spectrum of a porous Ti/SnO_2–Sb_2O_3/PbO_2 electrode. It can be seen that there are mainly Ti, Sn, Sb, O, Pb, and C peaks in the whole electrode. Figure 2b shows the Sn 3d spectrum with two characteristic peaks at 487.1 eV and 495.5 eV. Figure 2c shows the Sb 3d and O 1s spectra with characteristic peaks at 540.3 eV and 531.6 eV. Figure 2d shows the Pb 4f spectrum with characteristic peaks at 138.8 eV and 143.7 eV. The composition and chemical state of the porous Ti/SnO_2–Sb_2O_3/PbO_2 electrode could be determined by the XPS characterization results, and it was inferred that porous Ti/SnO_2–Sb_2O_3/PbO_2 electrode was successfully prepared.

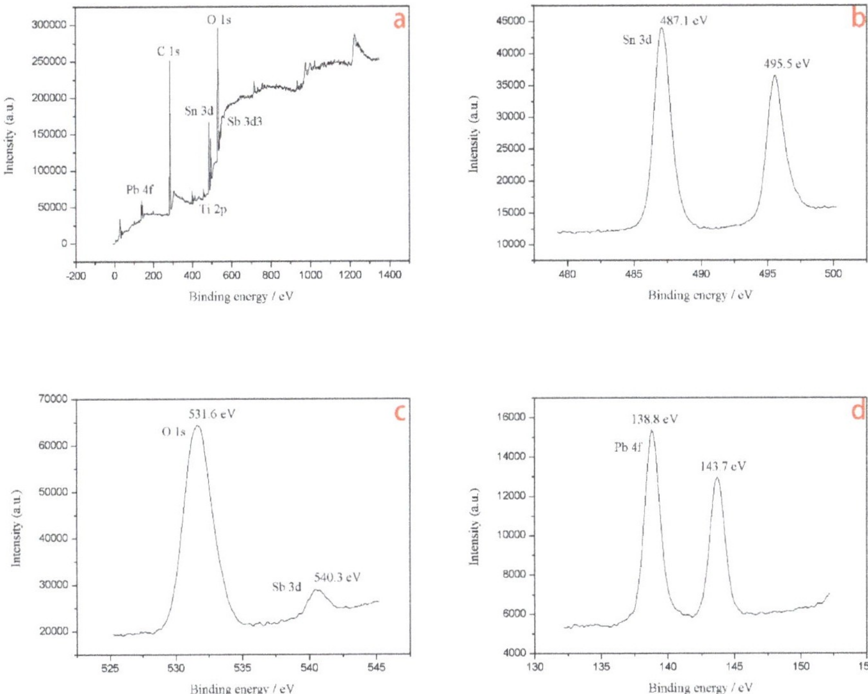

Figure 2. (**a**) XPS spectrum of the porous Ti/SnO_2–Sb_2O_3/PbO_2 electrode; (**b**) XPS spectrum of Sn 3d; (**c**) XPS spectrum of O 1s and Sb 3d; (**d**) XPS spectrum of Pb 4f.

3.1.3. X-ray Diffraction (XRD) Characterization

To further verify the results, the XRD patterns of the electrode coatings prepared on the three different-pore-size Ti substrates are shown in Figure 3. Since the XRD images of the three pore sizes electrodes are the same, only the 150 μm porous Ti/SnO_2–Sb_2O_3/PbO_2 electrodes are analyzed here. Figure 3 shows the β–PbO_2 diffraction peaks, which are (110), (101), (200), (211), (220), (310), (301), (321), (312), and (411). At the same time, Figure 3 shows the weak α–PbO_2 diffraction peaks, which is (041). In addition, there are no diffraction peaks of Ti, SnO_2, and Sb_2O_3, which proves that the PbO_2 coating had completely covered the porous Ti/SnO_2–Sb_2O_3 electrodes.

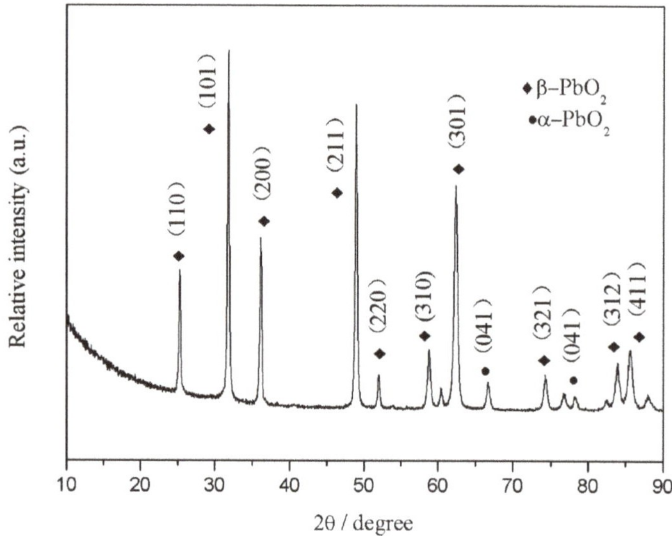

Figure 3. The X-ray diffraction (XRD) pattern of 150 μm porous $Ti/SnO_2–Sb_2O_3/PbO_2$ electrode.

3.2. Electrochemical Characterization

3.2.1. Linear Sweep Voltammetry (LSV) Curves

Figure 4 shows the linear sweep voltammetry (LSV) curves of the three different kinds of porous $Ti/SnO_2–Sb_2O_3/PbO_2$ electrodes as obtained in a 0.5 mol L^{-1} H_2SO_4 solution at a scan rate of 5 mV s^{-1}. The result shows that the oxygen evolution potential is 2.02 V, 1.99 V, and 1.96 V at the 150 μm, 100 μm, and 50 μm porous electrodes, respectively. High oxygen evolution potentials indicate that side reactions involving the formation of oxidized species are not very likely to occur [19]. 150 μm porous $Ti/SnO_2–Sb_2O_3/PbO_2$ electrodes therefore have the best removal effect on chloride ions.

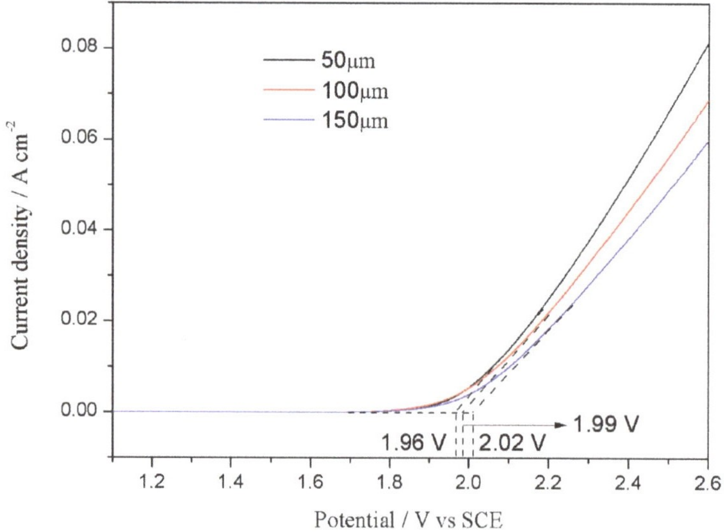

Figure 4. Linear sweep voltammetry (LSV) curves of porous $Ti/SnO_2–Sb_2O_3/PbO_2$ electrodes in 0.5 mol L^{-1} H_2SO_4 solution at a scan rate of 5 mV s^{-1}.

3.2.2. Cyclic Voltammetry

Figure 5 shows the cyclic voltammetry of the three different-pore-size Ti/SnO_2–Sb_2O_3/PbO_2 electrodes in 0.5 mol L^{-1} H_2SO_4 solution at a scan rate of 5 mV s^{-1}. The results show that the redox peak of the 50 μm porous Ti/SnO_2–Sb_2O_3/PbO_2 electrode is highest, while it is lowest at the 150 μm porous Ti/SnO_2–Sb_2O_3/PbO_2 electrode. The low oxygen evolution potential of the 50 μm porous Ti/SnO_2–Sb_2O_3/PbO_2 electrode indicates that it has a high areal density of oxygen evolution sites. The high oxygen evolution potential of the 150 μm porous Ti/SnO_2–Sb_2O_3/PbO_2 electrode, on the other hand, indicates a low areal density of oxygen evolution sites, which is beneficial for the electrocatalytic oxidation of chloride ions [20].

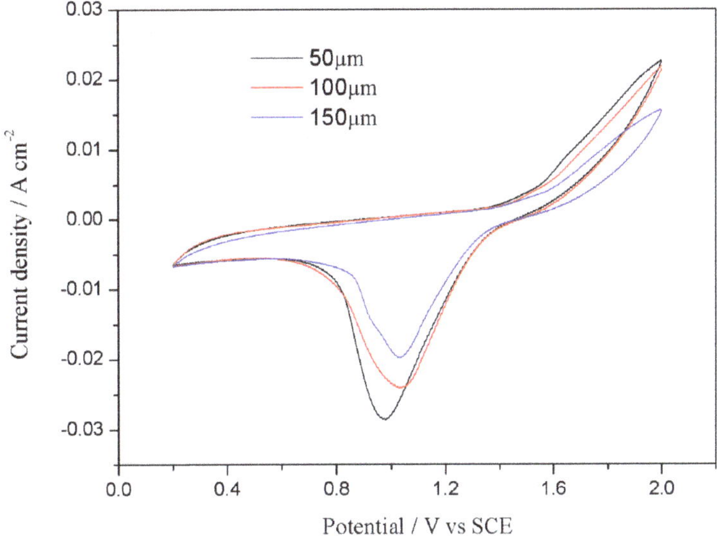

Figure 5. Cyclic voltammetry curve of porous Ti/SnO_2–Sb_2O_3/PbO_2 electrodes in 0.5 mol L^{-1} H_2SO_4 solution at a scan rate of 5 mV s^{-1}.

3.2.3. Linear Sweep Voltammetry (LSV) Curves in Different pH Solutions

Figure 6 shows the LSV curves of a 150 μm porous Ti/SnO_2–Sb_2O_3/PbO_2 electrode in different pH solutions at a scan rate of 5 mV s^{-1}. Solutions with pH values between pH = 3 and pH = 5 were prepared from buffer solution of citric acid and sodium citrate, and solutions with pH = 7 from buffer solution of potassium monohydrogen phosphate and potassium dihydrogen phosphate. Solutions with pH values between pH = 9 and pH = 11, finally, were composed from sodium carbonate and sodium bicarbonate buffer solution. It can be seen from the Figure 6 that under the same voltage conditions, the current density is smallest at pH = 9, which indicates that oxygen evolution is least likely to occur under these conditions and that therefore 150 μm porous Ti/SnO_2–Sb_2O_3/PbO_2 electrodes will exhibit an optimum chloride ion removal effect at pH = 9.

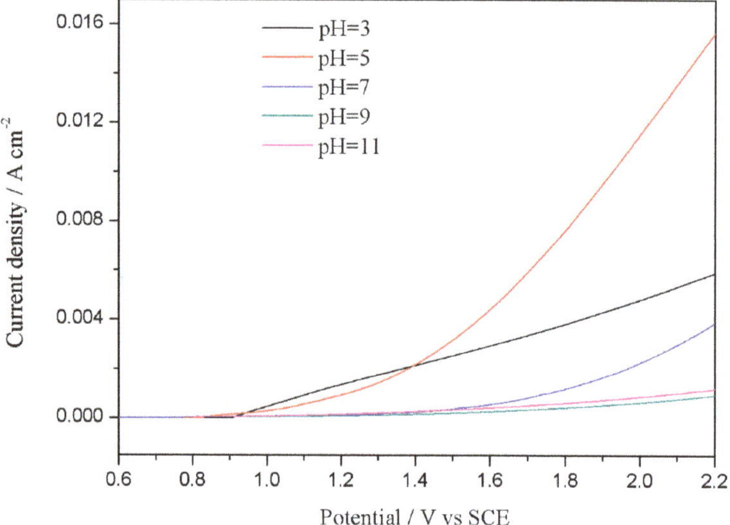

Figure 6. Linear sweep voltammetry (LSV) curves in different pH solutions of the 150 μm porous Ti/SnO_2–Sb_2O_3/PbO_2 electrode in 0.5 mol L^{-1} H_2SO_4 solution at a scan rate of 5 mV s^{-1}.

3.3. Electrocatalytic Oxidation of Chloride Ions

3.3.1. Effect of Pore Size

With the electrolysis time fixed at 4 h, the electrode plate spacing at 1.0 cm, the NaCl concentration at 10 g L^{-1}, and the current density at 100 mA cm^{-2}, the chloride ion removal ratios of the electrodes were determined to explore the optimal electrode pore size among three different kinds of porous electrodes in NaCl solution. As shown in Figure 7a, the optimal pore size of porous Ti/SnO_2-Sb_2O_3/PbO_2 electrode is 150 μm after 4 h, and the removal ratio of chloride ions is 92.8%, which confirms that the 150 μm pore size electrode performs best among the above three different kinds of porous electrodes. In addition, the effect of removing chloride ions from the 150 μm porous Ti/SnO_2-Sb_2O_3 electrode and porous Ti/SnO_2–Sb_2O_3/PbO_2 electrode is also compared. It can be seen from Figure 7b that the porous electrode with PbO_2 coating has a much better effect on removing chloride ions than the electrode without PbO_2 coating. The reason for this improvement is that the stability and the oxygen evolution potential of the porous Ti/SnO_2–Sb_2O_3/PbO_2electrode are further improved by the addition of the PbO_2 coating, which is beneficial for achieving an increased removal ratio of chloride ions.

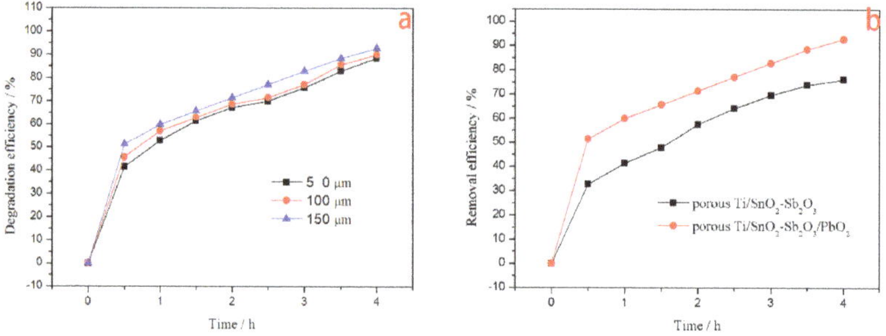

Figure 7. (**a**) Effect of pore size on chloride ion removal efficiency; (**b**) chloride ion removal efficiency as observed with 150 μm porous Ti/SnO_2-Sb_2O_3 and 150 μm porous Ti/SnO_2-Sb_2O_3/PbO_2 electrodes.

3.3.2. Effect of Initial NaCl Concentration

Related studies have shown that the initial ion concentration can affect the chloride ion removal ratio and cell voltage. As shown in Figure 8, with the electrolysis time fixed at 4 h, the plate spacing at 1.0 cm, and the current density at 100 mA cm^{-2}, the removal ratio of chloride ions and the cell voltage both decrease as the NaCl concentration is increased. The reason for this effect is that the conductivity is increased and the electrical resistance decreased as the NaCl concentration is increased [1]. In addition, it could be seen from the figure that the 150 μm porous Ti/SnO_2–Sb_2O_3/PbO_2 electrode has a superior removal effect on chloride ions at low concentrations. Therefore, considering the combined effects of the removal ratio chloride ions and the cell voltage, 10 g L^{-1} NaCl solution were selected as the following study.

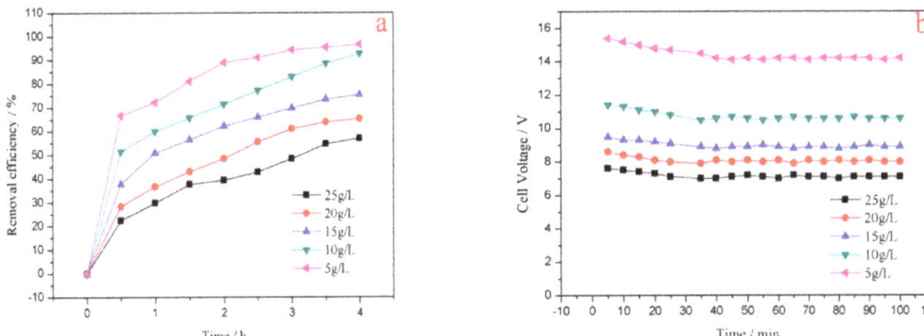

Figure 8. (**a**) Chloride ions removal efficiency under various initial NaCl concentration; (**b**) change in cell voltage under various initial NaCl concentrations.

3.3.3. Effect of Current Density

With the electrolysis time at 4 h, the optimal reaction current density was determined for 150 μm porous Ti/SnO_2–Sb_2O_3/PbO_2 electrodes. As shown in Figure 9, with a 150 μm electrode, a NaCl concentration 10 g L^{-1}, and the electrode plate spacing 1.0 cm, the removal ratio of chloride ions almost reached an upper limit at 125 mA cm^{-2} within 4 h. Current densities higher than 125 mA cm^{-2}, tended to damage the electrodes, resulting in reduced electrode lifetimes. So, it could be concluded that 125 mA cm^{-2} is the optimal current density.

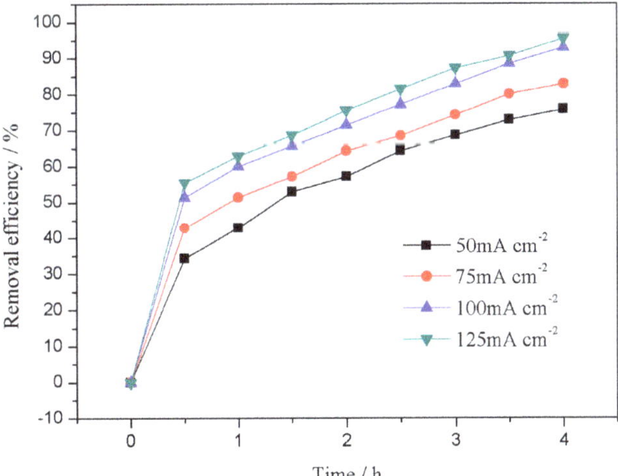

Figure 9. Chloride ion removal efficiency under various current densities.

3.3.4. Effect of Initial pH

As can be seen in Figure 10, with the 150 μm electrode, the NaCl concentration 10 g L^{-1}, the electrode plate spacing 1.0 cm, and the current density 125 mA cm^{-2}, the removal ratio of chloride ions is shown under the pH from 3 to 11. The solution of pH = 3 and pH = 5 was prepared by adding diluted concentrated sulfuric acid, and the solution of pH = 9 and pH = 11 was prepared by adding a diluted sodium hydroxide solution. It can be seen from the figure when the initial pH is the weak acidity, weak alkali, or neutral conditions, and the removal ratio of chloride ions is very effective, the removal ratio could reach 98.5% at pH = 9. In addition, the porous Ti/SnO_2–Sb_2O_3/PbO_2 electrode has a good removal effect over the whole range of pH values.

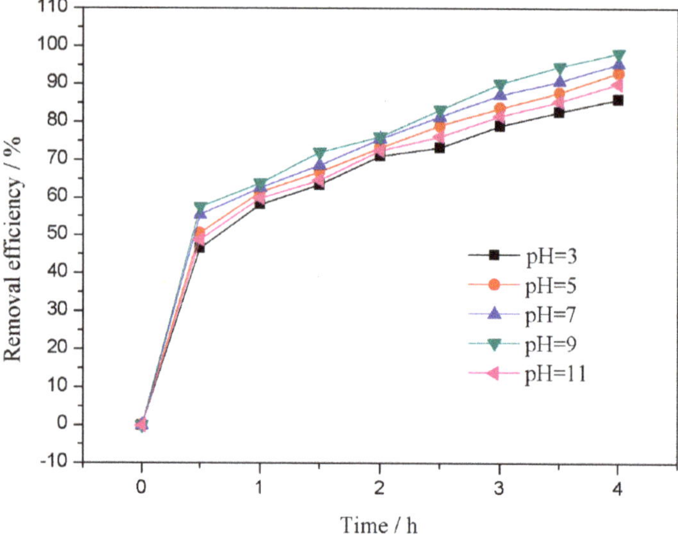

Figure 10. Chloride ion removal efficiency as a function of initial pH value.

3.3.5. Effect of Electrode Plate Spacing

Through the previous work, it was found that a change in the spacing between cathode and anode plates would cause a change in the cell voltage, which will impact the energy consumption of the whole process. As can be seen from Figure 11, the cell voltage at a current density of 125 mA cm^{-2} is about 11.7 V, when the spacing between both the plates is 1.0 cm and when the 150 μm electrode is operated in a NaCl solution with a concentration of 10 g L^{-1} and at an initial pH = 9. Reduction of the plate spacing to 0.75 cm lowers the cell voltage to about 10.4 V; a further reduction to 9.4 V occurs when the plate spacing is reduced to 0.5 cm. Therefore, it can be seen that the voltage decreases as the spacing between both plates is reduced. A reduction in the spacing of the anode and cathode plates therefore can lower the cell voltage and thereby the power consumption.

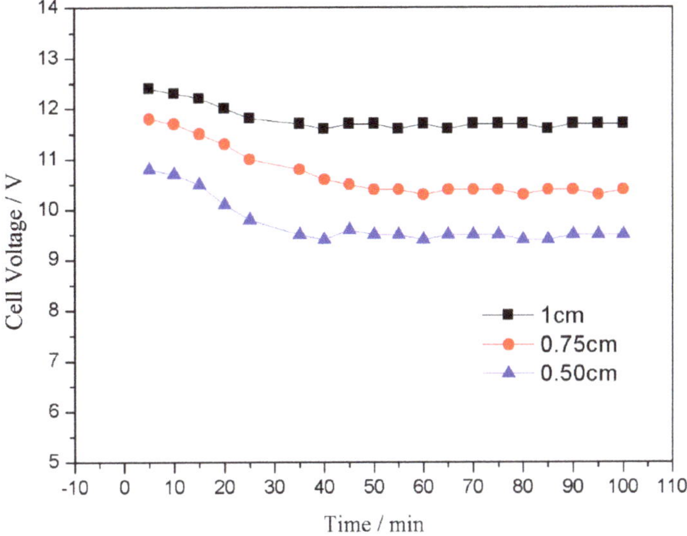

Figure 11. Change in cell voltage with electrode plate spacing.

3.3.6. Number of Cycles

With the 150 μm electrode, a NaCl concentration of 10 g L^{-1}, an initial pH = 9, a current density 125 mA cm^{-2}, and the spacing between plates 0.5 cm, Figure 12a shows the change in removal ratio of chloride ions of the porous electrode after repeating the removal process under optimal conditions for ten times. It can be seen that the removal ratio of chloride ions decreases after 10 times. However, the removal ratio of chloride ions of the porous Ti/SnO_2–Sb_2O_3/PbO_2 electrode is still more than 90%, which indicates that the porous Ti/SnO_2–Sb_2O_3/PbO_2 electrode has great advantages in stability. In addition, Figure 12b,c show the SEM and XRD patterns of the electrode after recycling, Figure 12c shows that the XRD spectrum of the electrode has not changed with the previous electrode after recycling, and Figure 12b shows that the PbO_2 deposition coating of the electrode has suffered partial damage after recycling. By zooming 4000 times, it is found that the PbO_2 deposition coating has developed many cracks, and thus it may be suspected that the reduction in removal rate might be related to these cracks.

Figure 12. (**a**) Change in chloride removal rate of a porous Ti/SnO_2–Sb_2O_3/PbO_2 electrode during ten successive removal cycles; (**b**) Scanning electron microscopy (SEM) pattern of the porous Ti/SnO_2–Sb_2O_3/PbO_2 electrode after recycling; (**c**) X-ray diffraction (XRD) pattern of the porous Ti/SnO_2–Sb_2O_3/PbO_2 electrode after recycling.

3.3.7. Mechanism of Removing Chloride Ions

During the experiment, the sodium chloride solution was poured into the electrolytic cell. The chlorine evolution reaction and other side reactions mainly occurred at the anode. The hydrogen evolution reaction and other side reactions mainly occurred at the cathode. The chloride ions formed chlorine gas at the anode and the produced chlorine gas was absorbed in the absorption tank, which in turn resulted in the removal of chloride ions.

Anode reaction:

$$2Cl^- - 2e^- \rightarrow Cl_2 \uparrow \tag{1}$$

$$4OH^- - 4e^- \rightarrow 2H_2O + O_2 \uparrow \tag{2}$$

Cathode reaction:

$$2H_2O + 2e^- \rightarrow 2OH^- + H_2 \uparrow \tag{3}$$

$$ClO^- + 2H_2O + 2e^- \rightarrow 2OH^- + Cl^- \tag{4}$$

Side reaction in solution:

$$Cl_2 + 2OH^- \rightarrow ClO^- + Cl^- + H_2O \tag{5}$$

$$Cl_2 + H_2O \rightarrow HClO + H^+ + Cl^- \tag{6}$$

$$HClO + H_2O \rightarrow H^+ + ClO^- \tag{7}$$

It can be seen from the above reaction formulae that the side reaction (2) competes with the desired chloride removal reaction (1) for holes from the anode. Due to the high oxygen evolution potential of porous Ti/SnO_2–Sb_2O_3/PbO_2 electrodes, reaction (2), however, is inhibited. With reaction (2) being inhibited, the generation of OH^- ions at the cathode via reaction (3) will also be inhibited. Under alkaline conditions, side reactions (5)–(7) will tend to produce ClO^- ions, which, in turn, will become converted into Cl^- ions via reaction (4) and which will finally become removed by the main reaction (1). As previous studies have shown that the concentrations of hypochlorite and perchlorate ions which are generated in the electrolysis of sodium chloride solutions are actually quite small [21], side reactions (5)–(7) are not expected to make any major contribution to the overall chlorine removal process [22,23]. Overall, our considerations therefore reveal that porous Ti/SnO_2–Sb_2O_3/PbO_2 electrodes can efficiently remove chloride ions from aqueous solutions, thus opening very broad application prospects.

4. Conclusions

In this paper, 150 µm porous Ti/SnO_2–Sb_2O_3/PbO_2 electrodes have been demonstrated to exhibit a good removal effect on chloride ions from NaCl solutions via the process of electrocatalytic oxidation. Factors that were found to influence the process of electrocatalytic oxidation were initial concentration, pore size, current density, initial pH, electrode plate spacing, and the number of removal cycles. Removal ratios of chloride ions up to 98.5% were observed under the following conditions: initial concentration 10 g L^{-1}, current density 125 mA cm^{-2}, electrode plate pore size 150 µm, initial pH = 9, and electrode plate spacing 0.5 cm. Under these conditions, the porous electrodes exhibited good stability. Physicochemical and electrochemical characterization results showed that porous Ti/SnO_2–Sb_2O_3/PbO_2 electrodes had a high oxygen evolution potential. Overall, it appears that porous Ti/SnO_2–Sb_2O_3/PbO_2 electrodes can play an important role in the electrocatalytic oxidation of chloride ions in water, thus opening prospects for a wide range of applications.

Author Contributions: Conceptualization, K.X.; Data curation, P.C.; Formal analysis, W.G.; Project administration, P.Y.; Resources, Y.L.; Writing—review & editing, J.P.

Funding: The authors declare no funding.

Acknowledgments: The study was supported by the college of chemistry and molecular sciences at Wuhan University.

Conflicts of Interest: The authors declare no conflict of interest.

References

1. Cui, L.; Li, G.P.; Li, Y.Z.; Yang, B.; Zhang, L.Q.; Dong, Y.; Ma, C.Y. Electrolysis-electrodialysis process for removing chloride ion in wet flue gas desulfurization wastewater (DW): Influencing factors and energy consumption analysis. *Chem. Eng. Res. Des.* **2017**, *123*, 240–247. [CrossRef]
2. Abdel-Wahab, A.; Batchelor, B. Chloride removal from recycled cooling water using ultra-high lime with aluminum process. *Water Environ. Res.* **2002**, *74*, 256–263. [CrossRef] [PubMed]
3. Woolard, C.R.; Irvine, R.L. Biological treatment of hypersaline wastewater by a biofilm of halophific bacteria. *Water Environ. Res.* **1994**, *66*, 230–235. [CrossRef]
4. Vyrides, I.; Stuckey, D.C. Saline sewage treatment using a submerged anaerobic membrane reactor (SAMBR): Effects of activated carbon addition and biogas-sparging time. *Water Res.* **2009**, *43*, 933–942. [CrossRef] [PubMed]
5. Walha, K.; Amar, R.B.; Firdaous, L.; Quéméneur, F.; Jaouen, P. Brackish groundwater treatment by nanofiltration, reverse osmosis and electrodialysis in tunisia: Performance and cost comparison. *Desalination* **2007**, *207*, 95–106. [CrossRef]
6. Mansoor, K. New nanopore zeolite membranes for water treatment. *Desalination* **2010**, *251*, 176–180.
7. Shaw, W.A. Fundamentals of zero liquid discharge system design. *Power* **2011**, *155*, 56–63.
8. Lv, L.; Sun, P.D.; Gu, Z.Y.; Du, H.G.; Pang, X.J.; Tao, X.H.; Xu, R.F.; Xu, L.L. Removal of chloride ion from aqueous solution by ZnAl-NO_3 layered double hydroxides as anion-exchanger. *J. Hazard. Mater.* **2009**, *161*, 1444–1449. [CrossRef]
9. Li, H.S.; Chen, Y.H.; Long, J.Y.; Jiang, D.Q.; Liu, J.; Li, S.J.; Qi, J.Y.; Zhang, P.; Wang, J.; Gong, J.; et al. Simultaneous removal of thallium and chloride from a highly saline industrial wastewater using modified anion exchange resins. *J. Hazard. Mater.* **2017**, *333*, 179–185. [CrossRef]
10. Jiang, S.X.; Li, Y.N.; Bradley, P.L. A review of reverse osmosis membrane fouling and control strategies. *Sci. Total Environ.* **2017**, *595*, 567–583. [CrossRef]
11. Li, D.; Tang, J.Y.; Zhou, X.Z.; Li, J.S.; Sun, X.Y.; Shen, J.Y.; Wang, L.J.; Han, W.Q. Electrochemical degradation of pyridine by Ti/SnO_2-Sb tubular porous electrode. *Chemosphere* **2016**, *149*, 49–56. [CrossRef] [PubMed]
12. Yang, K.; Lin, H.; Liang, S.T.; Xie, R.Z.; Lv, S.H.; Niu, J.F.; Chen, J.; Hu, Y.Y. A reactive electrochemical filter system with an excellent penetration flux porous Ti/SnO_2-Sb filter for efficient contaminant removal from water. *RSC Adv.* **2018**, *8*, 13933–13944. [CrossRef]
13. An, H.; Li, Q.; Tao, D.J.; Cui, H.; Xu, X.T.; Ding, L.; Sun, L.; Zhai, J.P. The synthesis and characterization of Ti/SnO_2-Sb_2O_3/PbO_2 electrodes: The influence of morphology caused by different electrochemical deposition time. *Appl. Surf. Sci.* **2011**, *258*, 218–224. [CrossRef]
14. Ding, H.Y.; Feng, Y.J.; Liu, J.F. Preparation and properties of Ti/SnO_2-Sb_2O_5 electrodes by electrodeposition. *Mater. Lett.* **2007**, *61*, 4920–4923. [CrossRef]
15. Zhao, W.; Xing, J.T.; Chen, D.H.; Bai, Z.L.; Xia, Y.S. Study on the performance of an improved Ti/SnO_2-Sb_2O_3/PbO_2 based on porous titanium substrate compared with planar titanium substrate. *RSC Adv.* **2015**, *5*, 26530–26539. [CrossRef]
16. Zhao, W.; Xing, J.T.; Chen, D.H.; Jin, D.Y.; Shen, J. Electrochemical degradation of Musk ketone in aqueous solutions using a novel porous Ti/SnO_2-Sb_2O_3/PbO_2 electrodes. *J. Electroanal. Chem.* **2016**, *775*, 179–199. [CrossRef]
17. Xing, J.T.; Chen, D.H.; Zhao, W.; Peng, X.L.; Bai, Z.L.; Zhang, W.W.; Zhao, X.X. Preparation and characterization of a novel porous Ti/SnO_2-Sb_2O_3-CNT/PbO_2 electrode for anodic oxidation of phenol wastewater. *RSC Adv.* **2015**, *5*, 53504–53513. [CrossRef]
18. Sun, J.R.; Lu, H.Y.; Lin, H.B.; Huang, W.M.; Li, H.D.; Lu, J.; Cui, T. Boron doped diamond electrodes based on porous Ti substrate. *Mater. Lett.* **2012**, *83*, 112–114. [CrossRef]

19. Ishibashi, K.; Fujishima, A.; Watanabe, T.; Hashimoto, K. Detection of active oxidative species in TiO_2 photocatalysis using the fluorescence technique. *Electrochem. Commun.* **2000**, *2*, 207–210. [CrossRef]
20. Zhang, W.L.; Lin, H.B.; Kong, H.S.; Lu, H.Y.; Yang, Z.; Liu, T.T. High energy density PbO_2/activated carbon asymmetric electrochemical capacitor based on lead dioxide electrode with three-dimensional porous titanium substrate. *Int. J. Hydrogen Energy* **2014**, *39*, 17153–17161. [CrossRef]
21. Neodo, S.; Rosestolato, D.; Ferro, S.; De Battisti, A. On the electrolysis of dilute chloride solutions: Influence of the electrode material on Faradaic efficiency for active chlorine, chlorate and perchlorate. *Electrochim. Acta* **2012**, *80*, 282–291. [CrossRef]
22. Duirk, S.E.; Desetto, L.M.; Davis, G.M. Transformation of Organophosphorus Pesticides in the Presence of Aqueous Chlorine: Kinetics, Pathways, and Structure–Activity Relationships. *Environ. Sci. Technol.* **2009**, *43*, 2335–2340. [CrossRef] [PubMed]
23. Cherney, D.P.; Duirk, S.E.; Tarr, J.C.; Collette, T.W. Monitoring the speciation of aqueous free chlorine from pH 1 to 12 with Raman spectroscopy to determine the identity of the potent low-pH oxidant. *Appl. Spectrosc.* **2006**, *60*, 764–772. [CrossRef] [PubMed]

Article

A Novel Porous Ni, Ce-Doped PbO_2 Electrode for Efficient Treatment of Chloride Ion in Wastewater

Sheng Liu, Lin Gui, Ruichao Peng and Ping Yu *

College of Chemistry & Molecular Science, Wuhan University, Wuhan 430072, China; sgg520@whu.edu.cn (S.L.); lingui@whu.edu.cn (L.G.); prc@whu.edu.cn (R.P.)

* Correspondence: yuping@whu.edu.cn; Tel.: +86-027-6875-2511

Received: 29 March 2020; Accepted: 9 April 2020; Published: 16 April 2020

Abstract: The porous Ti/Sb-SnO_2/Ni-Ce-PbO_2 electrode was prepared by using a porous Ti plate as a substrate, an Sb-doped SnO_2 as an intermediate, and a PbO_2 doped with Ni and Ce as an active layer. The surface morphology and crystal structure of the electrode were characterized by scanning electron microscope(SEM), energy dispersive spectrometer(EDS), and X-Ray diffraction(XRD). The electrochemical performance of the electrodes was tested by linear sweep voltammetry (LSV), cyclic voltammetry (CV), electrochemical impedance spectroscopy (EIS), and electrode life test. The results show that the novel porous Ni-Ce-PbO_2 electrodes with larger active surface area have better electrochemical activity and longer electrode life than porous undoped PbO_2 electrodes and flat Ni-Ce-PbO_2 electrodes. In this work, the removal of Cl^- in simulated wastewater on three electrodes was also studied. The results show that the removal effect of the porous Ni-Ce-PbO_2 electrode is obviously better than the other two electrodes, and the removal rate is 87.4%, while the removal rates of the other two electrodes were 72.90% and 80.20%, respectively. In addition, the mechanism of electrochemical dechlorinating was also studied. With the progress of electrolysis, we find that the increase of OH^- inhibits the degradation of Cl^-, however, the porous Ni-Ce-PbO_2 electrode can effectively improve the removal of Cl^-.

Keywords: porous Ni-Ce-PbO_2; co-doping; active surface area; removal rate

1. Introduction

The widespread use of chlorinated compounds such as HCl, NaCl, and $MgCl_2$ in the industrial field has increased the content of chloride ion in wastewater [1,2]. If it is discharged into the water body beyond control, the water environment will be seriously damaged. The accumulation of chloride ions will make the soil salinized and alkalized, and excessive intake of chlorine by the human body will cause organ damage. Chloride ions are corrosive to pipelines, boilers, etc., and can erode buildings and reduce durability of the concrete structure. For example, a large amount of Cl^- in desulfurization wastewater discharged from thermal power plants can corrode pipelines and equipments [3].

At present, the most widely used method for treating Cl^- in wastewater is chemical precipitation [4]. However, the concentration of Cl^- in the treated wastewater is still high. The treatment of chloride ion in wastewater by chemical precipitation will be limited in the future [5]. How to treat desulfurization wastewater in depth, meet the discharge requirements, and reduce the impact on the environment has always been a difficult problem in the field of wastewater treatment at home and abroad. Therefore, advanced oxidation processes (AOPs) [6,7] such as catalytic ozonation, Fenton oxidation, supercritical water oxidation, electrochemical oxidation, photocatalytic oxidation, etc., [8] have been studied. Electrochemical oxidation is one of the most eye-catching AOPs. It has the advantages of good treatment effect, small floor area, high degradation efficiency, short residence time and no secondary pollution, and has broad application prospects in the advanced treatment of salt compounds and organic compounds [9]. In the recent years, electrochemical oxidation technology has been increasingly studied for the treatment of chloride ion in wastewater [10].

The degradation efficiency and degradation products of electrochemical oxidation process change with the anode material [11], which means that anode material is one of the main factors in electrochemical oxidation process [12,13]. Because of the critical role of anode materials, scholars have been studying new anodes for many years [14]. The earliest graphite and carbon electrodes have the disadvantages of low current efficiency and poor mechanical strength [15]. Subsequently, metal anodes were developed. At present, metal oxide electrode is widely used and its preparation process is mature, such as dimensionally stable anode (DSA) (e.g., RuO_2, IrO_2, PbO_2, SnO_2) [11]. With the deepening of research, 3D porous structure compounds are widely used in the preparation of anode materials because of their large specific surface area, such as carbon nanotubes (CNT), porous graphene (GE), etc., [16,17], which can significantly change the structure of the coating when doped in the metal oxide coating [11]. Scholars have found that doping metal elements in metal oxide coatings can significantly improve the electrode activity and extend service life [18], such as Ce, Bi, Fe, Co, etc., [11]. In addition, it has been found that many rare earth oxides doped into PbO_2 can significantly improve the electrochemical performance of the electrode [19]. Kong et al. reported that doping Er_2O_3, Gd_2O_3, La_2O_3, and Ce_2O_3 on PbO_2 electrode could promote the degradation of 4-chlorophenol [20]. Jin et al. [20] found that doping Ce can form smaller crystal size on the surface of the electrode, resulting in the increase of specific surface area and catalytically active sites. Xia et al. [21] reported that proper doping of Ni can make the grains dense, which not only facilitates electron movement, and improves electrochemical performance, but also helps to extend electrode life.

In this work, we intended to use a porous Ti plate as the substrate, because the porous structure has large specific surface area. We can improve the electrochemical activity and stability of the electrode by doping Ce and Ni in the active layer PbO_2 electrode [22]. The doping of Ce and Ni reduces the crystal size of the active layer [11,21], but increases the surface area of the active layer and the electrochemical activity of the electrode, in addition, the service life of the electrode is extended to some extent. Finally, the electrochemical activity and stability of porous Ni-Ce-PbO_2 electrodes were investigated and compared with the conventional porous undoped PbO_2 electrode and the doped flat Ni-Ce-PbO_2 electrode. In this experiment, simulated wastewater with high chlorine content was selected as the target pollutant. The electrochemical performance of the electrode was investigated by comparing the effects of three kinds of electrodes on the treatment of chloride ions in simulated wastewater.

2. Experimental

2.1. Materials

Porous titanium plate with a purity of 99.9% was purchased from Baoji Jinkai Technology Co., Ltd., Jinan, China. $Ni(NO_3)_2\ 6H_2O$ was purchased from Xiqiao Chemical Co. Ltd., Foshan, China. All other chemicals were purchased from Sino pharm Chemical Reagent Co. Ltd., Shangai, China. All chemicals were of analytical grade and used as received. In this work, deionized water was used in all solutions.

2.2. Electrode Preparation

2.2.1. Titanium Surface Preparation

The porous titanium plates (20 mm × 10 mm × 2.8 mm) bought by Baoji Jinkai Technology Co., Ltd., Jinan, China, had a purity of 99.9% and an average pore diameter of 50 μm to pretreat the substrate. The porous titanium plate was ultrasonically cleaned in acetone for 15 min, washed in 20% NaOH solution at 90 °C for 1 h, and etched in a 15 wt % oxalic acid solution at 90 °C for 1 h until a gray matte titanium matrix was formed, and finally saved in ethanol.

The surface treatment of the flat titanium plate (20 mm × 10 mm × 2.8 mm) is similar to porous titanium plate.

2.2.2. Coating SnO_2–Sb_2O_3

The electrode intermediate layer was prepared by thermal decomposition method [11]. Total of 1.2 g $SnCl_4$, 0.2 g Sb_2O_3, and 10 mL concentrated hydrochloric acid were dissolved in 25 mL of isopropanol to obtain a precursor-coating solution. The solution was colorless, transparent, slightly sticky. Then the pretreated porous titanium plate was immersed in the precursor solution. After soaking, it was taken out and dried in an oven at 110 °C, and then calcined in a muffle furnace at 500 °C for 15 min. After cooling the plate, the drying and calcining process was repeated several times, and the high-temperature baking time in the last time was extended to 1 h to obtain a porous Ti/Sb-SnO_2 electrode. The main purpose of this layer is to improve the conductivity of the electrode and prevent the titanium matrix from being oxidized to form TiO_2.

The precursor solution can be directly brushed on the surface of the flat titanium plate with a brush. Other preparation processes of the intermediate layer of the flat electrode are similar to that of the porous electrode.

2.2.3. Electrochemical Deposition Ni–Ce–PbO_2

$Pb(NO_3)_2$, $Ni(NO_3)_2$, and $Ce(NO_3)_2$ were dissolved in 250 mL water at a ratio of 100:1:1 of Pb, Ni, and Ce. Then, 0.04 M NaF and 4 mL/L of PTFE were added, before adding 0.1 mol/L HNO_3 to adjust the pH to 1, to form an electrodeposition solution. The control temperature was 65 °C, the current density was 20 mA/cm^2, the electrodeposition time was 1 h, so as to deposit a surface layer of lead dioxide with Ce, Ni co-doped on the surface of the intermediate layer tin antimony oxide, that is, porous Ti/Sb-SnO_2/Ni-Ce-PbO_2 electrode.

In addition to the raw materials (Ni $(NO_3)_2$ and Ce $(NO_3)_2$), the preparation process of porous PbO_2 electrode is similar to that of porous Ni-Ce PbO_2 electrode

2.3. Electrode Characterization

The surface morphology was observed using a scanning electron microscope (Quanta 200 of FEI, Hillsboro, OR, USA). X-ray diffraction (XRD) patterns of samples were obtained with an X-ray diffractometer (PANalytical, Almelo, The Netherlands). Cyclic voltammetry (CV), linear sweep voltammetry (LSV), and electrochemical impedance spectroscopy (EIS) were performed at room temperature using a computer-controlled electrochemical workstation (CHI 660E, CH Instruments, Shanghai, China) with a conventional three-electrode system. The prepared PbO_2-based electrode (20 mm × 20 mm) was used as the working electrode, a saturated Ag/AgCl electrode was employed as the reference electrode, and a stainless steel sheet was applied as the counter electrode. All potentials were referred to the SCE. The stability tests (up to 20 h) were performed by the accelerated life test with a current density of 1 A·cm^{-2} and a temperature of 60 °C in 2 M H_2SO_4 solution for porous undoped PbO_2 electrodes, porous Ni-Ce–PbO_2 electrodes, and Flat Ni-Ce–PbO_2 electrodes. These tests were performed in a three-electrode system.

2.4. Electrochemical Oxidation

A simulated chlorine-containing wastewater with a chloride ion concentration of 4 g/L and a volume of 200 mL was selected as the experimental wastewater; the temperature was 298 K in all experiments. The electrochemical oxidation experiment was conducted by a batch method, and the device was mainly composed of a DC power source, a collector types magnetic stirrer, and a glass reactor. The anode (PbO_2-based electrodes) and the cathode (flat titanium plate) were placed parallel to each other and perpendicular to the solution level with a distance of 1 cm. The volume of simulated wastewater in all experiments was 200 mL and the Cl^- concentration was 4 g/L. The temperature of all experiments was maintained at 298 K. All pH values in the experiment were determined by a pH meter and all Cl^- concentration in the experiment were determined by titration with a standard $AgNO_3$ solution.

3. Results and Discussion

3.1. Surface Morphological and Crystallographic Analysis

The SEM images of the plate-modified Ni-Ce-PbO_2 electrode, the porous undoped PbO_2 electrode, and the porous Ni-Ce-PbO_2 electrode are shown in Figure 1. It can be seen that the porous undoped PbO_2 electrode and the porous Ni-Ce-PbO_2 electrode still have many small holes in the electrode surface while compared with porous Ti plate (Figure 1a), indicating that the PbO_2 coating did not block the porous structure. From Figure 1b,f it can be seen that the porous Ni-Ce-PbO_2 electrodes apparently have larger pores and specific surface area compared to the flat Ni-Ce-PbO_2 electrodes.

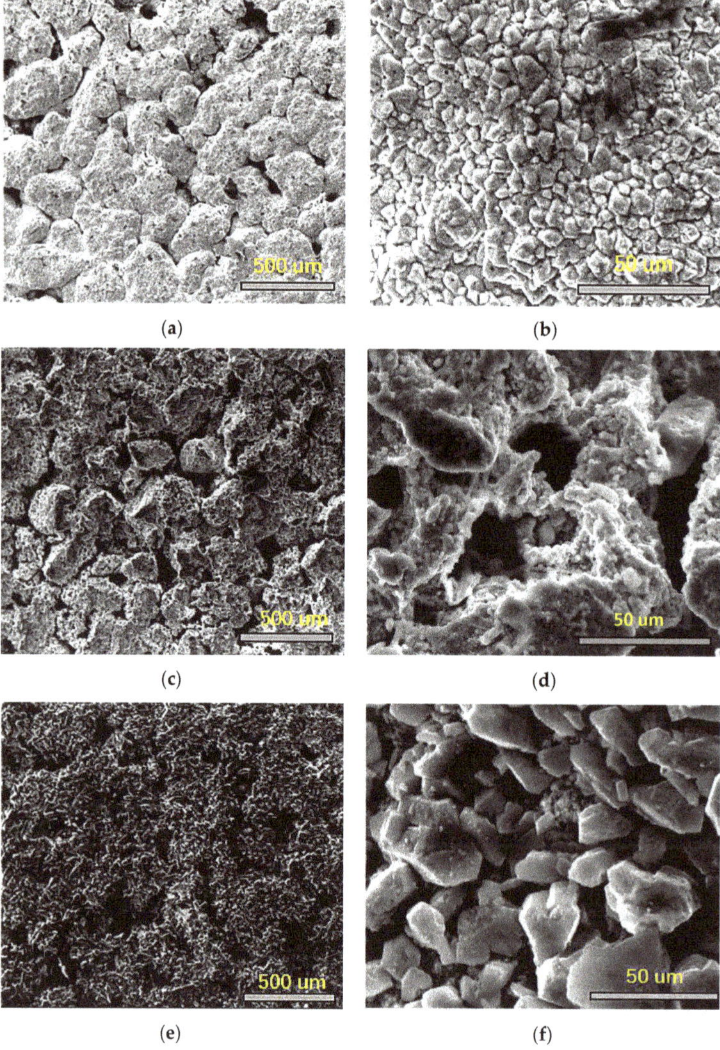

Figure 1. SEM of different electrodes, (**a**) (porous titanium plate, 80×); (**b**) (flat Ni-Ce-PbO_2 electrode, 1200×) and (**c**) (porous undoped PbO_2 electrode, 80×); (**d**) (porous undoped PbO_2 electrode, 1200×) and (**e**) (porous Ni-Ce-PbO_2 electrode, 80×); (**f**) (porous Ni-Ce-PbO_2 electrode, 1200×).

The porous PbO_2 electrode doped with Ni and Ce (Figure 1e,f) can make the holes smaller or bigger on the surface of the electrodes, and the specific surface area was larger than porous undoped PbO_2 electrodes (Figure 1c,d). Besides, the EDS spectrum of different PbO_2 electrodes is shown in Figure 2a, which confirms that there are O, Pb, Ce, Ni elements in the porous Ni-Ce-PbO_2 electrode, while there are only O and Pb elements in the porous undoped PbO_2 electrodes. Thus, it can be concluded that the Ni and Ce was successfully doped into the PbO_2 films and doping of Ni and Ce can reduce the grain size of the PbO_2 coating.

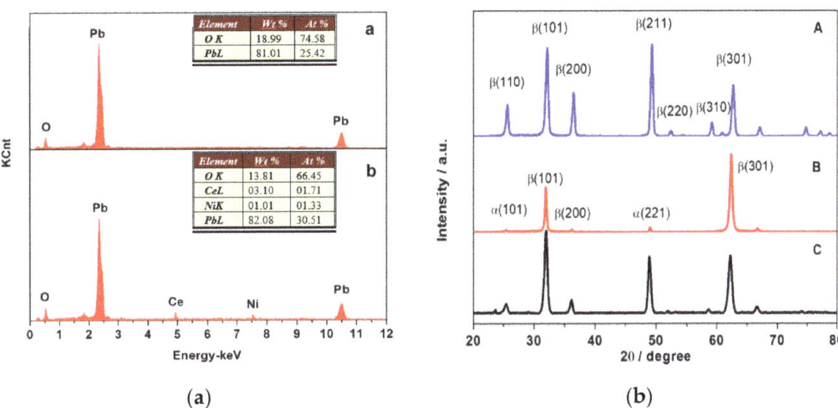

(**a**) (**b**)

Figure 2. (**a**) EDS of a (porous PbO_2 electrode) and b (porous Ni-Ce-PbO_2 electrode); (**b**) XRD diffraction patterns of different Electrodes A (Porous Ti/Sb-SnO_2/PbO_2 electrode), B (Porous Ti/Sb-SnO_2/Ni-Ce-PbO_2 electrode), and C (Standard XRD pattern of PbO_2).

From Figure 2b, it can be seen that the diffraction peaks of undoped porous show that dioxide electrodes appear at 25.4 degrees, 31.9 degrees, 36.1 degrees, 48.9 degrees, 58.8 degrees, 62.4 degrees, and 66.8 degrees, which are consistent with the JCPDS card (41–1492) in pattern, indicating that the main component of undoped PbO_2 surface layer is β-PbO_2 and the coating of porous Ni-Ce-PbO_2 electrode consists of a mixture of crystalline phases of α and β-PbO_2. The content of α-PbO_2 is higher than that of undoped PbO_2 because the doping of Ce changes the preferred crystalline orientation of the electrode surface and forms smaller grains. In addition, compared with the undoped PbO_2, the intensity of the diffraction peaks of Ni-Ce-PbO_2 decreases or even disappears because of the doping of Ni and Ce. The reason may be that doping of Ni and Ce changes the nucleation and growth of crystals in the coating, making the electrode have smaller crystal size than the undoped PbO_2 electrode, which can also be seen from the SEM image. The average crystallite size calculated from the width of [101] diffraction peaks by Scherrer's formula is 17.64 nm (Ce–Ni) and 28.76 nm (undoped). According to literature [20], the diffraction peak width is inversely proportional to the crystallite size. The result indicates that the deposited Ni-Ce-PbO_2 has smaller crystallite size than other electrode. Smaller crystal size may help to form larger specific surface area which may lead to better electrochemical performance.

3.2. Electrochemical Performance Test

As shown in Figure 3a, no redox peak signal was observed on any of the electrodes in the blank Na_2SO_4 solution, indicating that PbO_2 is an electrochemically inert material in the blank Na_2SO_4 solution. After the addition of Cl^- (Figure 3b), a distinct oxidation peak was obtained. There is no doubt that the oxidation peak is attributed to the oxidation of Cl^- on the surface of the anode. However, no corresponding reduction peak was observed in the reverse scanning from 3 V to 0 V, indicating that the oxidation of Cl^- is a completely irreversible electrode reaction process. The oxidation peak potential of the porous Ni-Ce-PbO_2 electrode (2.01 V vs. SCE) was lower than that of the other two electrodes

(2.24 and 2.48 V), but the oxidation peaks current (0.031 A) was significantly higher than the other two electrodes (0.028 A and 0.024 A), which showed that the porous Ni-Ce-PbO_2 electrode has higher electro-catalytic activity for Cl^-, and the improvement of its electro-catalytic activity is not only related to the increase in electrode surface area caused by its porous structure and doping with Ni and Ce, but also due to changes in the PbO_2 band structure. The electrode doping Ni and Ce not only increases the donor area of PbO_2, but also increases the donor level of PbO_2, making it easier for electrons to jump from the donor level to the conduction band [21]. Therefore, the conductivity of PbO_2 is improved.

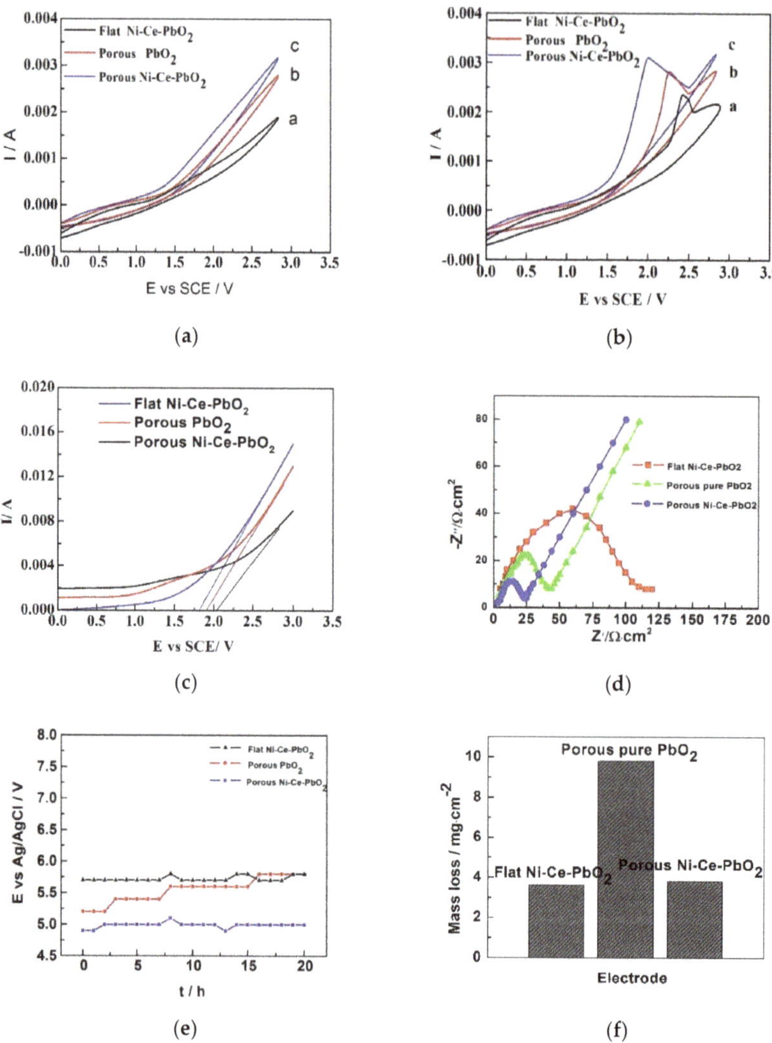

Figure 3. (**a**) Cyclic voltammetry (CV) of different PbO_2 electrodes measured in 0.1mol·L^{-1} Na_2SO_4 solution, (**b**) 0.1 mol·L^{-1} Na_2SO_4 solution (pH = 6.5) containing 4 g L^{-1} Cl^-, scan rate: 50 mV s^{-1}, T = 298 K, (**c**) LSV curves of different electrodes measured in 0.5 M Na_2SO_4, scan rate: 10 mV s^{-1}, T = 298 K, (**d**) EIS of different electrodes. Conditions: T = 298 K; [H_2SO_4] = 1 M. (**e**) Electrode stability tests: electrode potential vs. time for the electrolysis using different electrodes. (**f**) The mass losses of electrodes after accelerated life tests for 20 h.

The oxygen evolution overpotential (OEP) of different electrodes could be measured by LSV. According to the polarization curve in Figure 3c, the OEP of the porous Ni-Ce-PbO_2 electrode was the highest with 2.09 V (vs. SCE) compared with the flat Ni-Ce-PbO_2 electrode of 1.81 V (vs. SCE) and the porous PbO_2 electrode of 1.91 V (vs. SCE), respectively. The electrodes with high OEP values can produce more hydroxyl radicals [23]. In addition, it can be seen from Figure 3c that the porous Ni-Ce-PbO_2 electrodes had higher oxygen evolution current than the other two electrodes. The higher oxygen releases current also verified that the active surface area of the porous Ni-Ce-PbO_2 electrode is larger than that of the other PbO_2-based electrodes.

In Figure 3d, there is an obvious semicircle in the electrochemical impedance spectroscopy (EIS) of the three electrodes. The radius of the semicircle usually reflects the magnitude of the transfer resistance R_{ct} of the chlorine evolution or oxygen evolution reaction, which is available from the EIS spectrum. The R_{ct} sizes of Ni-Ce-PbO_2 electrodes, porous PbO_2 electrodes, and porous Ni-Ce-PbO_2 electrodes were 59.4, 21.2, and 12.2, respectively. It is indicated that the porous Ni-Ce-PbO_2 electrode has the best conductivity, the most active surface sites, and the highest chlorine evolution activity.

The service life is a critical factor in the practical application of the electrode. Under the condition of anode current of 1A pa^{-2} and temperature of 60 °C, the accelerated life test of different PbO_2 electrodes was carried out in 2M H_2SO_4 solution for 20 h. According to Figure 3e,f, the Ni-Ce-PbO_2 electrode exhibited better electrochemical stability and less mass loss than the pure PbO_2 electrode. It is well-known that the main reasons of mass loss of electrode are the separation and dissolution of PbO_2 film, and the mass loss is proportional to the service life of the electrode. Therefore, it can be concluded that the stability of Ni-Ce-PbO_2 electrode is much better than that of pure PbO_2 electrode by the modification of Ni and Ce. The mass loss of the porous Ni-Ce-PbO_2 electrode is slightly larger than that of the flat Ni-Ce-PbO_2, mainly because of the loose porosity of the structure. In fact, in the experiments, we found that film peeled off on the pure PbO_2 electrode, but not on the other two modified PbO_2 electrodes.

3.3. Electrochemical Oxidation of Cl^- in Simulated Wastewater

The electrochemical oxidation of simulated wastewater was compared with three kinds of electrodes. From Figure 4a, after electrolysis for 100 min, the removal rate of Cl^- on the porous Ni-Ce-PbO_2 electrode was as high as 87.4%. At the same time, the plate Ni-Ce-PbO_2 electrode and the porous PbO_2 electrode were used as anodes, and the highest removal rates of Cl^- were 72.90% and 80.20%, respectively. Obviously, the above results show that the electrochemical oxidation ability of the porous Ni-Ce-PbO_2 electrode was much higher than the other two electrodes. This may be due to the fact that the porous structure increases the surface area of the electrode and the doping of Ni and Ce further increases the surface area of the electrode, which facilitates the adsorption of Cl^- on the surface of the anode and promotes the mass transfer and exchange of the reactants. Thus, the electrochemical oxidation ability of the electrode is improved. As is shown in Figure 4b that in the early stage of electrolysis, the removal rate of Cl^- was higher than that in the later stage because of the enrichment of Cl^- in wastewater. The results show that higher current density result in higher Cl^- removal rate, and the difference in Cl^- removal rate was not obvious at relatively higher current densities. With the progress of electrolysis, the decrease of Cl^- concentration inhibited the oxidation of Cl^- and reached equilibrium at a certain time, in which Cl^- cannot be completely removed. However, the porous Ni-Ce-PbO_2 electrode with large specific surface area can effectively improve the removal rate of Cl^-. During the process of degradation of Cl^-, the chlorine evolution reaction and oxygen evolution side reaction in the anode proceed simultaneously, the content of Cl^- is higher and the content of OH^- is lower in the initial stage of reaction process (Figure 4c). The reaction formulas are as follows:

$$\text{Anode: } 2Cl^- \rightarrow Cl_2 + 2e^- \quad (1)$$

$$Cl_2 + 2H_2O \rightarrow HClO + H_3O^+ + Cl^- \quad (2)$$

$$HClO + H_2O \rightarrow H_3O^+ + ClO^- \tag{3}$$

$$\text{Cathode: } H_2O \rightarrow H^+ + OH^- \tag{4}$$

$$2H^+ + 2e^- \rightarrow H_2 \tag{5}$$

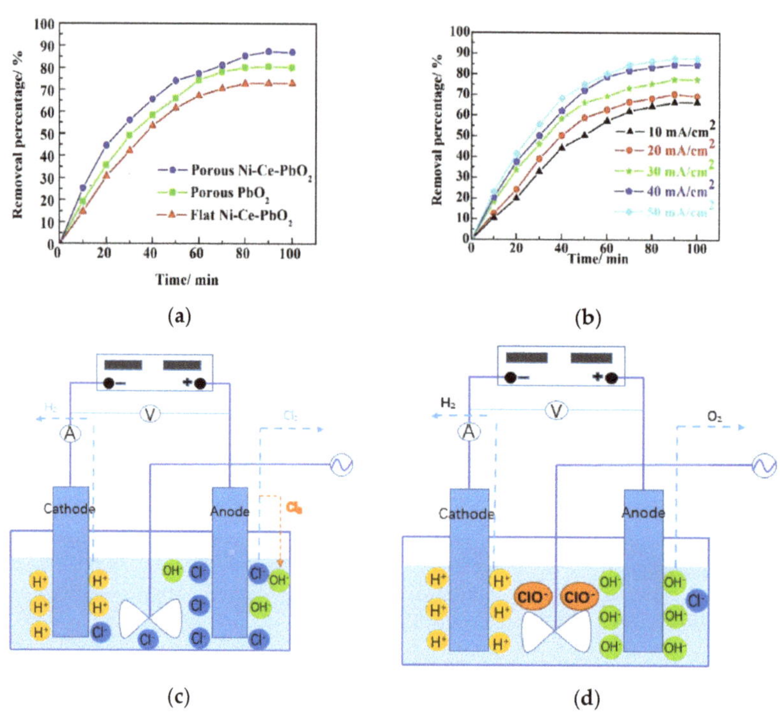

Figure 4. (**a**) Variation of Cl^- removal percentage with time during electrochemical oxidation on different electrodes, conditions: current density = 50 mA/cm^2; T = 298 K; $[Cl^-]$ = 4 g/L, (**b**) at different current densities, conditions: porous Ni-Ce-PbO_2 electrode; T = 298 K; $[Cl^-]$ = 4 g/L and electrochemical reaction process of electrode dechlorinating (**c**) at 30 min and (**d**) at 100 min.

According to the investigation [24], under the experimental conditions, the anode mainly generates Cl_2 (Equation (1)). The cathode produces a large amount of OH^-, which exists in the solution to raise the pH of the solution quickly, except a small portion is transferred to the anode for consumption. The increase of OH^- concentration promotes the disproportionation of chlorine and forms a large number of ClO^- (Equations (2) and (3)). In addition, Figure 4a also shows the fact that the Cl^- removal rate increases significantly with the increase of the initial concentration. But after the electrolysis time reaches 90 min, a large amount of Cl^- is removed and the OH^- concentration in the solution is gradually increased. This can also be seen from Figure 4c,d, the reaction formulas are as follows:

$$\begin{gathered} H_2O \rightarrow H^+ + OH^- \\ \text{Anode: } 2OH^- - 4e^- \rightarrow O_2 + 2H^+ \\ \text{Cathode: } 4H^+ + 4e^- \rightarrow 2H_2 \end{gathered} \tag{6}$$

A large amount of OH groups accumulates on the surface of the anode, which hinders the contact between a small amount of Cl^- in the solution and the anode, so that the oxygen evolution reaction on the surface of the anode gradually becomes the main reaction. The consumption of OH^- on the anode results in a small part of residual Cl^- in the solution which cannot be removed, and another small part of Cl^- is converted into ClO^- and HClO [24] (Equations (1)–(3)), which greatly limits the removal of Cl^-. Also as a DSA electrode, the porous Ni-Ce-PbO_2 electrode degraded 87.4% of Cl^- at 90 min because of its large active surface area, while the degradation rate of the plated Ni-Ce-PbO_2 electrode and the porous PbO_2 electrode were only 72.90% and 80.20%, respectively. Therefore, the porous Ni-Ce-PbO_2 electrode breaks through the limitation of the conventional electrode in removing Cl^-. In addition, we also detected the concentration of Pb^{2+} and Ni^{2+} in the solution after the 5th oxidation, and found that the concentration of Pb^{2+} was only about 0.0085 mg/L, and Ni^{2+} was even less, far lower than the standard of the World Health Organization. It shows that it is environment friendly and will not cause secondary pollution.

4. Conclusions

In this paper, porous Ni-Ce-doped PbO_2 electrodes were successfully prepared on a porous titanium substrate by thermal deposition and electrodeposition. The surface morphology, crystal structure, electrochemical activity, and electrode stability of the flat Ni-Ce-PbO_2, porous undoped PbO_2, and porous Ni-Ce-PbO_2 were tested, and the electro-catalytic properties of these three electrodes in simulated wastewater were compared.

The porous Ni-Ce-PbO_2 electrodes possessed porous structure and smaller grain size than the other two electrodes. The doping of Ni and Ce can change the nucleation and growth of crystals on the surface of the electrode, making the electrode have smaller particles, larger electrochemical active surface area, and better electrode life.

At a current density of 50 mA, the porous Ni-Ce-PbO_2 electrode was used to treat Cl^- in the simulated wastewater. The removal rate was as high as 87.4%, while the highest removal rates of the porous pure PbO_2 electrode and the flat Ni-Ce-PbO_2 electrode were 72.90% and 80.20%, respectively.

In addition, in the later stage of Cl^- oxidation, because of the increase of pH of the electrolyte, oxygen evolution reaction mainly occurs in the anode, which results in a part of Cl^- that cannot be removed from the solution and the other part that dissolves in the solution in the form of ClO^- and HClO; removal rate of Cl^- is restricted. The novel porous Ni-Ce-PbO_2 electrodes can effectively improve the removal of Cl^- because of its greater electrochemically active surface area.

Author Contributions: S.L. and P.Y. conceived and designed the experiments; S.L., L.G., and R.P. performed the experiments; S.L. analyzed the data and wrote the paper; S.L. and P.Y. revised the manuscript. All authors have read and agreed to the published version of the manuscript.

Funding: This research received no external funding.

Conflicts of Interest: The authors declare no conflict of interest.

References

1. Guàrdia, M.D.; Guerrero, L.; Gelabert, J.; Gou, P.; Arnau, J. Sensory characterisation and consumer acceptability of small calibre fermented sausages with 50% substitution of NaCl by mixtures of KCl and potassium lactate. *Meat Sci.* **2008**, *80*, 1225–1230. [CrossRef] [PubMed]
2. Xia, J.; Liu, Q.F.; Mao, J.H.; Qian, Z.H.; Jin, S.J.; Hu, J.Y.; Jin, W.L. Effect of environmental temperature on efficiency of electrochemical chloride removal from concrete. *Constr. Build. Mater.* **2018**, *193*, 189–195. [CrossRef]
3. Yang, B.; Chen, Z.; Zhang, M.; Zhang, H.; Zhang, X.; Pan, G.; Zou, J.; Xiong, Z. Effects of elevated atmospheric CO_2 concentration and temperature on the soil profile methane distribution and diffusion in rice-wheat rotation system. *J. Environ. Sci. (China)* **2015**, *32*, 62–71. [CrossRef] [PubMed]
4. Gupta, V.K.; Ali, I.; Saleh, T.A.; Nayak, A.; Agarwal, S. Chemical treatment technologies for waste-water recycling—An overview. *RSC Adv.* **2012**, *2*, 6380–6388. [CrossRef]
5. Hu, S.; Ding, S.F.; Fan, Z.S. Zero release technology of desulfurization waste water in coal—Fired power plant. *Clean Coal Technol.* **2015**, *21*, 129–133.
6. Liu, H.; Liu, Y.; Zhang, C.; Shen, R. Electrocatalytic oxidation of nitrophenols in aqueous solution using modified PbO_2 electrodes. *J. Appl. Electrochem.* **2008**, *38*, 101–108. [CrossRef]

7. Deng, Y.; Zhao, R. Advanced Oxidation Processes (AOPs) in Wastewater Treatment. *Curr. Pollut. Rep.* **2015**, *1*, 167–176. [CrossRef]
8. Shmychkova, O.; Luk'yanenko, T.; Velichenko, A.; Meda, L.; Amadelli, R. Bi-doped PbO_2 anodes: Electrodeposition and physico-chemical properties. *Electrochim. Acta* **2013**, *111*, 332–338. [CrossRef]
9. Xu, M.; Wang, Z.; Wang, F.; Hong, P.; Wang, C.; Ouyang, X.; Zhu, C.; Wei, Y.; Hun, Y.; Fang, W. Fabrication of cerium doped Ti/nanoTiO_2/PbO_2 electrode with improved electrocatalytic activity and its application in organic degradation. *Electrochim. Acta* **2016**, *201*, 240–250. [CrossRef]
10. Shuangchen, M.; Jin, C.; Gongda, C.; Weijing, Y.; Sijie, Z. Research on desulfurization wastewater evaporation: Present and future perspectives. *Renew. Sustain. Energy Rev.* **2016**, *58*, 1143–1151. [CrossRef]
11. Duan, X.; Zhao, Y.; Liu, W.; Chang, L.; Li, X. Electrochemical degradation of p-nitrophenol on carbon nanotube and Ce-modified-PbO_2 electrode. *J. Taiwan Inst. Chem. Eng.* **2014**, *45*, 2975–2985. [CrossRef]
12. Dai, Q.; Xia, Y.; Chen, J. Mechanism of enhanced electrochemical degradation of highly concentrated aspirin wastewater using a rare earth La-Y co-doped PbO_2 electrode. *Electrochim. Acta* **2016**, *188*, 871–881. [CrossRef]
13. Xia, Y.; Dai, Q.; Chen, J. Electrochemical degradation of aspirin using a Ni doped PbO_2 electrode. *J. Electroanal. Chem.* **2015**, *744*, 117–125. [CrossRef]
14. Yao, Y.; Teng, G.; Yang, Y.; Huang, C.; Liu, B.; Guo, L. Electrochemical oxidation of acetamiprid using Yb-doped PbO_2 electrodes: Electrode characterization, influencing factors and degradation pathways. *Sep. Purif. Technol.* **2019**, *211*, 456–466. [CrossRef]
15. Elaissaoui, I.; Akrout, H.; Grassini, S.; Fulginiti, D.; Bousselmi, L. Effect of coating method on the structure and properties of a novel PbO_2 anode for electrochemical oxidation of Amaranth dye. *Chemosphere* **2019**, *217*, 26–34. [CrossRef] [PubMed]
16. Xu, F.; Chang, L.; Duan, X.; Bai, W.; Sui, X.; Zhao, X. A novel layer-by-layer CNT/PbO_2 anode for high-efficiency removal of PCP-Na through combining adsorption/electrosorption and electrocatalysis. *Electrochim. Acta* **2019**, *300*, 53–66. [CrossRef]
17. Zhou, X.; Liu, S.; Yu, H.; Xu, A.; Li, J.; Sun, X.; Shen, J.; Han, W.; Wang, L. Electrochemical oxidation of pyrrole, pyrazole and tetrazole using a TiO_2 nanotubes based SnO_2-Sb/3D highly ordered macro-porous PbO_2 electrode. *J. Electroanal. Chem.* **2018**, *826*, 181–190. [CrossRef]
18. Santos, J.E.L.; de Moura, D.C.; da Silva, D.R. Application of TiO_2-nanotubes/PbO_2 as an anode for the electrochemical elimination of Acid Red 1 dye. *J. Solid State Electrochem.* **2018**, *23*, 351–360. [CrossRef]
19. Du, H.; Duan, G.; Wang, N.; Liu, J.; Tang, Y.; Pang, R.; Chen, Y.; Wan, P. Fabrication of Ga_2O_3–PbO_2 electrode and its performance in electrochemical advanced oxidation processes. *J. Solid State Electrochem.* **2018**, *22*, 3799–3806. [CrossRef]
20. Jin, Y.; Wang, F.; Xu, M.; Hun, Y.; Fang, W.; Wei, Y.; Zhu, C.G. Preparation and characterization of Ce and PVP co-doped PbO_2 electrode for waste water treatment. *J. Taiwan Inst. Chem. Eng.* **2015**, *51*, 135–142. [CrossRef]
21. Wang, Z.; Xu, M.; Wang, F.; Liang, X.; Wei, Y.; Hu, Y.; Zhu, C.G.; Fang, W. Preparation and characterization of a novel Ce doped PbO_2 electrode based on NiO modified Ti/TiO_2NTs substrate for the electrocatalytic degradation of phenol wastewater. *Electrochim. Acta* **2017**, *247*, 535–547. [CrossRef]
22. Yao, Y.; Huang, C.; Yang, Y.; Li, M.; Ren, B. Electrochemical removal of thiamethoxam using three-dimensional porous PbO_2-CeO_2 composite electrode: Electrode characterization, operational parameters optimization and degradation pathways. *Chem. Eng. J.* **2018**, *350*, 960–970. [CrossRef]
23. Xie, R.; Meng, X.; Sun, P.; Niu, J.; Jiang, W.; Bottomley, L.; Li, D.; Chen, Y.; Crittenden, J. Electrochemical oxidation of ofloxacin using a TiO_2-based SnO_2-Sb/polytetrafluoroethylene resin-PbO_2 electrode: Reaction kinetics and mass transfer impact. *Appl. Catal. B Environ.* **2017**, *203*, 515–525. [CrossRef]
24. Neodo, S.; Rosestolato, D.; Ferro, S.; De Battisti, A. On the electrolysis of dilute chloride solutions: Influence of the electrode material on Faradaic efficiency for active chlorine, chlorate and perchlorate. *Electrochim. Acta* **2012**, *80*, 282–291. [CrossRef]

Article

Experimental Study of Micro Electrochemical Discharge Machining of Ultra-Clear Glass with a Rotating Helical Tool

Yong Liu [1,*], Chao Zhang [2], Songsong Li [1], Chunsheng Guo [1,3] and Zhiyuan Wei [1]

1 Associated Engineering Research Center of Mechanics & Mechatronic Equipment, Shandong University, Weihai 264209, China; 201716276@mail.sdu.edu.cn (S.L.); 18369189101@163.com (Z.W.)
2 Department of Mechanical Engineering, Weihai Vocational Secondary School, Weihai 264213, China; zhangchao43711@163.com
3 Shenzhen Research Institute of Shandong University, Virtual University Park, Nanshan, Shenzhen 518057, China; guo@sdu.edu.cn
* Correspondence: rzliuyong@sdu.edu.cn; Tel.: +86-1356-312-3255

Received: 12 March 2019; Accepted: 29 March 2019; Published: 4 April 2019

Abstract: Electrochemical discharge machining (ECDM) is one effective way to fabricate non-conductive materials, such as quartz glass and ceramics. In this paper, the mathematical model for the machining process of ECDM was established. Then, sets of experiments were carried out to investigate the machining localization of ECDM with a rotating helical tool on ultra-clear glass. This paper discusses the effects of machining parameters including pulse voltage, duty factor, pulse frequency and feed rate on the side gap under different machining methods including electrochemical discharge drilling, electrochemical discharge milling and wire ECDM with a rotary helical tool. Finally, using the optimized parameters, ECDM with a rotary helical tool was a prospective method for machining micro holes, micro channels, micro slits, three-dimensional structures and complex closed structures with above ten micrometers side gaps on ultra-clear glass.

Keywords: electrochemical discharge machining; rotating helical tool; side gap; micro structures; closed structure; ultra-clear glass

1. Introduction

In recent years, ECDM has gained attention. Micro electromechanical systems (MEMS), including micro reactors and micro medical devices, often consist of the micro structures of nonconductive materials, such as glass, ceramics and silicon nitride. Therefore, the traditional machining method is difficult to use to fabricate micro structures composed of brittle and hard nonconductive materials. However, ECDM can machine micro structures on hard nonconductive materials. ECDM is a hybrid machining method including electrochemical machining and electric discharge machining [1]. When discharge takes place between the tool electrode and the surrounding electrolyte, local high temperatures and chemical reactions remove the workpiece material. Micro structures have been widely applied to micro accelerometers, micro pumps, micro containers and biological medical instruments, which could be machined by electrochemical machining (ECM), electro discharge machining (EDM) or ECDM [2–5]. Glass has superior properties, including transparency, high oxidation resistance, wear resistance, biological compatibility, and low electrical conductivity properties. ECDM, with a different machining method, can fabricate complex micro structures on glass, such as micro holes, micro channels, micro slits and complicated three-dimensional features.

ECDM was first put forward by Kurafuji in 1968. Because of machining brittle and hard nonconductive material at the micro level, this machining method was investigated further. Nasim,

Mohammad researched the generation of single gas bubbles at the tool electrode surface. Finally, he found that the wettability of the tool electrode and the surface tension between the bubble and electrolyte affected the gas film thickness [6]. Zhang and Huang explored critical voltage under different machining conditions by using ECDM on glass to investigate the time it took to form the gas film, via the mean current of discharge. They concluded that better machining precision and surface quality could be obtained by selecting optimized parameters [7]. Sathisha proposed the empirical model for the process of machining grooves with multiple regression analysis [8]. Jawalkar fabricated micro channels by electrochemical discharge milling. The experiment results showed that voltage plays a leading role in the parameters of both material removal rate and tool wear [9]. Cao found a new method indicating that the grinding process under polycrystalline diamond tools reduced the surface roughness of ECDM structures from a few tens of a micron to 0.05 μm Ra [10]. Elhami utilized special equipment to generate only a single discharge in ultrasonic-assisted electrochemical discharge machining (UAECDM) and studied two important characteristics: material removal and tool wear [11]. Many scholars conducted further studies of UAECDM [12–14]. Han and Min proposed a method of using the side insulation tool and low concentration electrolytes to reduce undesirable over cutting [15]. Furutani concluded that the width, depth and surface roughness of grooves machined by electrochemical discharge milling increased with higher voltage [16]. Kun investigated the precision and stability of quartz fabricated by ECDM and explored optimal machining parameters including the size of the electrode and the machining speed [17].

Wire electrochemical discharge machining (WECDM) was proposed by Tsuchiya [18]. Jain utilized traveling wire as a tool in WECDM, and studied the effects of voltage, the concentration of the electrolyte on material removal rate and tool wear [19]. Panda and Yadava established a 3D finite element transient thermal model and predicted the temperature field and MRR in traveling wire electrochemical spark machining (TW-ECSM) [20]. Kuo found a new wire ECDM approach to machine quartz glass. In their experiments, electrolytes were supplied by titrated flow and the machining quality and efficiency were improved [21]. Wang studied the surface integrity of alumina machined by WECDM [22]. A host of literature proved that electrolyte circulation plays an important role in machining performance. Many approaches to enhancing the electrolyte circulation in ECDM and wire ECDM have been proposed [23–25]. Fang used rotary helical electrodes in wire ECDM, which accelerated the cycle of the electrolyte [26]. Wang and Zhang researched the flow field of ECDM with a rotating helical tool. In their experiments, the gas–liquid phase distribution and the velocity vectors of the electrolyte in the machining gap were investigated [27].

In this paper, a rotating helical electrode was used in different ECDM processes, including electrochemical discharge drilling, electrochemical discharge milling and wire ECDM. The rotary helical electrode produced an axial velocity and axial force, dragging the electrolyte from the bottom of the workpiece. Therefore, the machining accuracy of ECDM is fine. The machining parameters, including voltage, frequency, duty factors, and feed rate, were considered in the experiments and their effect on the side gap was investigated. The optimized parameters were utilized to successfully machine micro holes, micro channels, micro slits and complicated three-dimensional features with ten several-micron side gaps.

2. Experimental Set-Up and Model for Machining Process

2.1. Experimental Set-Up

Most past efforts have been spent on studying the mechanisms of ECDM. The ECDM process can be depicted as in Figure 1. In all of the following experiments, the tool electrode with Φ105 μm was a rotating helical tungsten carbide (WC) electrode while the auxiliary anode is a graphite plate and the electrolyte was 3 mol/L KOH (Shuangshuang chemical industry, Yantai, China). This process can be divided into five steps. In the first step the pulse power was imposed on the tool electrode and the auxiliary anode, which were immersed in the electrolyte. Because of electrolysis, hydrogen and oxygen gas bubbles were generated around the tool electrode and auxiliary anode, respectively. The second step involved the hydrogen gas bubbles accumulating rapidly and embracing the tool electrode. The third step was when the formation rate of the hydrogen gas bubbles was equal to the rate of that escaping from the electrode. The gas film around the tool electrode was formed and completely separated the tool electrode from the surrounding electrolyte. In the fourth step there was a narrow gap between the tool electrode and the electrolyte according to the third step. When the applied voltage rose to a critical value, there was a spark in the gas film. As is known, a large amount of heat generated by discharge will instantaneously melt the surface material of the workpiece when the tool electrode is close to the workpiece. In addition, some material is removed due to evaporation and localized high temperature, leading KOH electrolytes to corrode the workpiece. The fifth step began when a gas film was staved when the tool electrode contacted with the electrolyte again. Then, the process switched back to the first step, beginning the cycle anew.

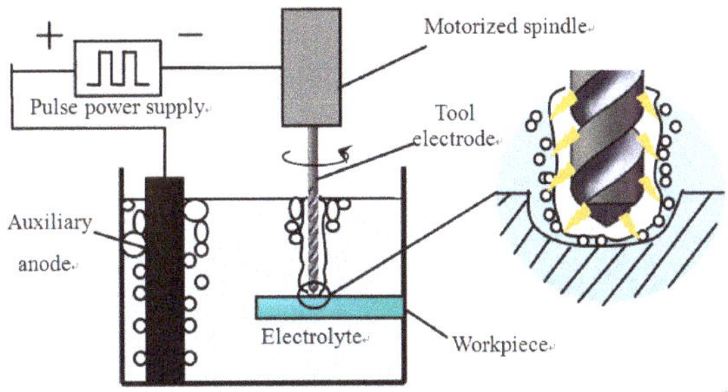

Figure 1. A schematic view of electrochemical discharge machining (ECDM).

The architecture of this experimental system for ECDM is illustrated in Figure 2. The experimental system contains four subsystems: the power supply system, machine tool system, microelectrode system, and processing control and monitoring system. The power supply system was plays a significant part in ECDM, which provides a series of variable ranges including pulse voltage, duty factor, and pulse frequency. The machine tool system is mainly comprised of an optical precision platform, the L shaped marble frame, feed device, high speed motorized spindle, lifting platform, fixture, and other components. The optical precision platform ensured high accuracy for micro ECDM. To guarantee the verticality of the machine tool, the L shaped marble frame possessing vibration isolation performance was used. The feed device, controlled by the MP-C154 motion control card, accurately controlled the feeding of the electric slipway along the three directions and met the requirements for fabricating complex three-dimensional micro structures. The microelectrode system consisted of a rotary helical WC electrode, electrolytic bath, high speed motorized spindle, fixture, and lifting platform. The glass workpiece was fixed on the electrolytic bath and placed on the lifting platform. The processing control and monitoring system had the motion control card and Supereyes. Supereyes monitored the process and captured images. In this research, electrochemical discharge drilling, electrochemical discharge milling, and wire ECDM were utilized to machine micro structures on glass workpieces with rotary helical WC electrodes via an experimental system.

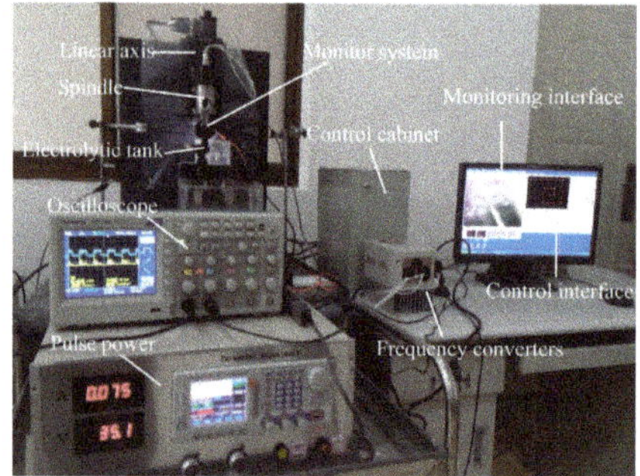

Figure 2. Experimental system of ECDM.

2.2. *Establishing of Machining Process Model*

To investigate the side gap in the ECDM process, three different types of experiments were carried out, including electrochemical discharge drilling, electrochemical discharge milling, and wire ECDM. During a certain specified experiment, only one parameter could be adjusted and the effect on the side gap recorder, all other parameters remained constant.

Step 1 was the model for electrochemical discharge drilling. Establishing the simplified model of electrochemical discharge drilling on the side gap needed the following hypothetical conditions:

(a) The mean heat released by the discharges q is linear to the energy for melting material in unit time, for which the linear coefficient is the constant k.
(b) The hole after drilling is a uniform cylinder.
(c) The distance between the end of the rotary helical electrode and the bottom of the hole is assumed to be constant and this constant is c, shown in Figure 3.

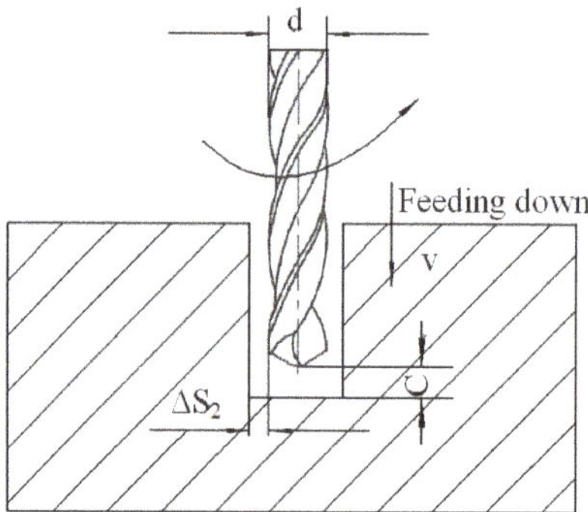

Figure 3. A schematic view of electrochemical discharge drilling.

The discharge energy q in unit time can be obtained by the equation proposed by Jain [28]:

$$q = UI - RI^2 \tag{1}$$

where U is voltage, I is the mean current, and R is the resistance between the cathode and the anode.

According to Assumption (a), the relationship between the discharge energy q and n is:

$$q = kn\lambda \tag{2}$$

where n is the amount of substance of melted glass in unit time and λ is the dissolution heat of ultra-clear glass.

The volume of melted glass, V, is worked out as

$$V = \frac{nM}{\rho}, \tag{3}$$

where M is the molar mass of the glass and ρ is the density of the glass. Therefore, the diameter, D, of the machined hole could be calculated together with Equation (3) as:

$$D = 2\sqrt{\frac{nM}{\pi h \rho}}, \tag{4}$$

where h is the drilling depth in unit time. The relationship between h and the feed rate v is:

$$h = v + c, \tag{5}$$

where v is the feed rate of the rotary helical electrode, c is the distance between the end of the rotary helical electrode and the bottom of the hole, according to Assumption (c).

The side gap ΔS_1 can be defined as follows, where the diameter of the rotary helical electrode is d:

$$\Delta S_1 = \frac{D-d}{2}. \tag{6}$$

The side gap ΔS_1 could be solved simultaneously with Equations (1), (2) and (4)–(6).

$$\Delta S_1 = \sqrt{\frac{M(UI - RI^2)}{\pi \rho k \lambda (v+c)}} - \frac{d}{2} \tag{7}$$

We concluded that side gap ΔS_1 rose with the increasing of the voltage, but decreased with higher feed rates. In addition, the side gap ΔS_1 was affected by material properties.

Step 2 was the model for electrochemical discharge milling and WECDM. The side gap was different between electrochemical discharge drilling and milling. The model of the side gap in electrochemical discharge milling ought to be reconstructed. The side gap model in the electrochemical discharge milling process is shown in Figure 4.

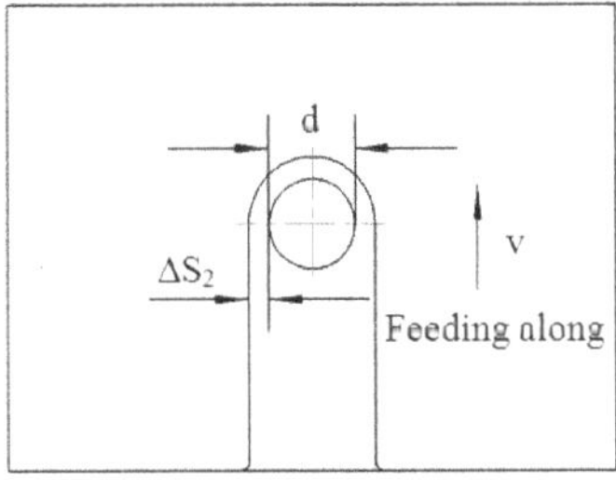

Figure 4. Schematic view of electrochemical discharge milling.

In unit time, the shape of the machined glass was considered rectangular in electrochemical discharge milling. Therefore, the volume of the machined glass could be obtained in unit time as follows:

$$V = (d + 2\Delta S_2) v h_1. \tag{8}$$

In Equation (8), v is the feed rate, h_1 is mean milling depth, and d is the diameter of the rotary helical electrode.

Therefore, the side gap was obtained by Equations (1)–(3) and (8).

$$\Delta S_2 = \frac{M(UI - RI^2)}{2\rho v h_1 k \lambda} - \frac{d}{2} \tag{9}$$

It was not hard to establish that the side gap ΔS_2 became larger with any increase of voltage in the electrochemical discharge milling. However, the side gap ΔS_2 became narrower with higher feed rate and higher milling depth. The glass properties also influenced the side gap.

The side gap in WECDM could be substituted, approximately, by Equation (9) from Figure 4, with an h_1 thickness of the glass.

3. Experiments and Discussion

3.1. Experimental Arrangement

In this paper, electrochemical discharge drilling, electrochemical discharge milling and wire ECDM were employed to investigate the side gap during the processing of ECDM. To ensure the accuracy of the experiments and to avoid accidental influence, each experiment was carried out repeatedly, at least three times. In all of the following experiments, Φ105 μm tungsten was used as the rotary helical tool and a 600 μm thick graphite plate was selected as the auxiliary electrode. Workpieces in the electrochemical discharge drilling and wire ECDM were ultra-glass with a thickness of 300 μm, while the specifications of the glass workpiece were 46 mm × 25 mm × 1 mm in electrochemical discharge milling. In addition, all feed depths were 100 μm in electrochemical discharge milling. The diameter and slit width were measured by a Nikon SMZ1270 microscope (Tokyo, Japan) and NOVA NANOSEM 450 scanning electron microscope (Hillsboro, OR, USA).

In these experiments, the auxiliary anode (Luhan metal, Shanghai, China), rotary helical electrode (Union tool, Tokyo Metropolitan, Janpan) and glass workpiece (Citoglas, Haimen, China) were immersed in electrolytes. When the pulse voltage was applied to the auxiliary anode and the helical electrode was attached to high speed spindle, a rotary helical electrode moved with a certain feed speed to machine the glass. The main discharge areas were the bottom, the side wall, and the side wall of the rotary helical electrode in the electrochemical discharge drilling, the electrochemical discharge milling, and the wire ECDM, respectively. Therefore, the selected experimental parameters were different between the three machining methods. The details of the experimental arrangements are shown in Table 1. In each group of experiments, only one parameter, the pulse voltage, pulse frequency, duty factor, or feed rate could be adjusted to the desirable range to research the effect on the side gap. Other variables were kept constant. The effects of the pulse voltage, frequency, duty factor, and feed rate on the side gap are displayed in the following table.

Table 1. The details of experimental arrangements.

Item	ECD-Drilling	ECD-Milling	Wire ECDM
Pulse voltage	35–41 (V)	34–40 (V)	32–40 (V)
Frequency	400–700 (Hz)	200–500 (Hz)	200–600 (Hz)
Duty factor	60–90 (%)	50–80 (%)	50–90 (%)
Feed velocity	0.5–2 (μm/s)	0.5–2 (μm/s)	0.5–2.5 (μm/s)
Spindle speed		3000 (rpm)	
Concentration		3 M (KOH)	

3.2. Effect of Pulse Voltage on Side Gap

There have been many experiments conducted to investigate effects of pulse voltage on the side gap. The side gap was calculated and the influence of the pulse voltage on the side gap is shown in Figure 5. From Figure 5 and Equations (7) and (9), we concluded that the side gap increased with the rise of the pulse voltage. At a lower pulse voltage, the bubbles generated by electrolysis were sparse and thin. Therefore, the thickness of the gas film was thin. The thin gas film and low applied voltage led to shorter discharge distances, which greatly shortened the side gap. While at higher pulse voltages, the formation rate of the bubbles increased rapidly. Plenty of bubbles coalesced intensely, resulting in a thicker gas film. Thus, in this case, the discharge distance was longer, meaning more material was removed. It was not difficult to conclude that the side gap increased with the rise of the discharge distance. The diameter of the hole in the electrochemical discharge drilling, the slit width in the electrochemical discharge milling and the wire ECDM increased with the higher pulse voltage.

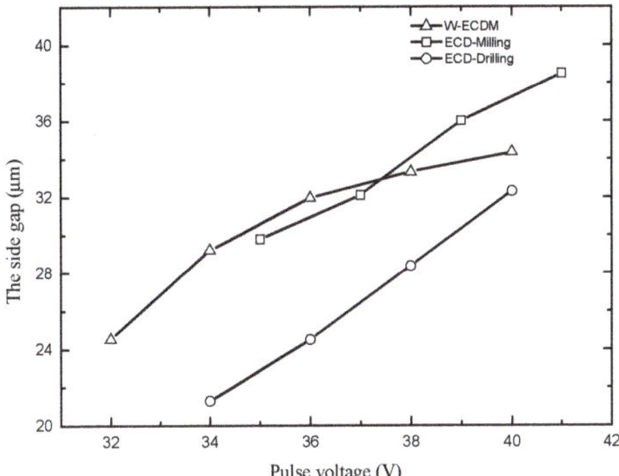

Figure 5. Effect of pulse voltage on side gap.

3.3. *Effect of Duty Factor on Side Gap*

To research the effect of the duty factor on the side gap, a series of experiments were carried out, including electrochemical discharge drilling, electrochemical discharge milling and wire ECDM. The results are shown in Figure 6. As the picture depicts, the side gap increases as the duty factor rises, from 40% to 90%. The discharge energy q in unit time increased due to the higher duty factor, which resulted in more material removal. Therefore, the diameter of the hole in electrochemical discharge drilling, the slit width in electrochemical discharge milling and the wire ECDM increased with the rise of the duty factor. The optimal duty factor should be low, but the lower duty factors reduced material removal rate.

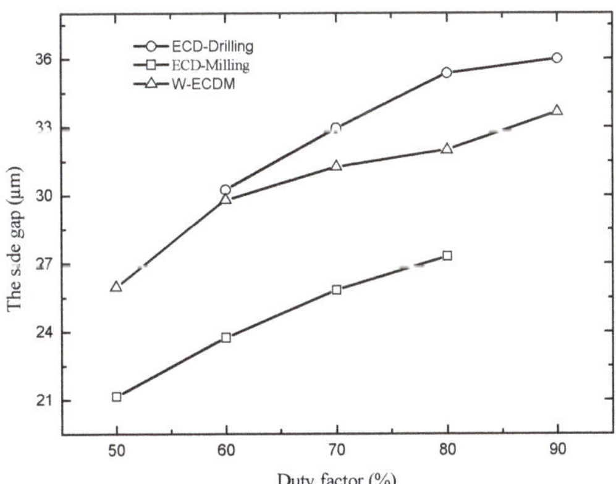

Figure 6. Effect of duty factor on side gap.

3.4. *Effect of Frequency on Side Gap*

The influence of frequency on the side gap is shown in Figure 7. In this set of experiments, the duty factor remained unchanged at 70% and frequency ranged from 200 Hz to 600 Hz. In electrochemical discharge drilling, electrochemical discharge milling, and wire ECDM the side gap decreased when the frequency increased, gradually. The number of discharge rose with higher frequency per unit time, but the pulse width was correspondingly reduced. Therefore, the discharge energy of a single discharge decreased, resulting in less material removal and smaller side gaps, eventually. The diameter of the hole in electrochemical discharge drilling, the slit width in electrochemical discharge milling and the wire ECDM decreased with the rise of frequency. The optimal frequency should be high but the higher frequency will reduce the material removal rate.

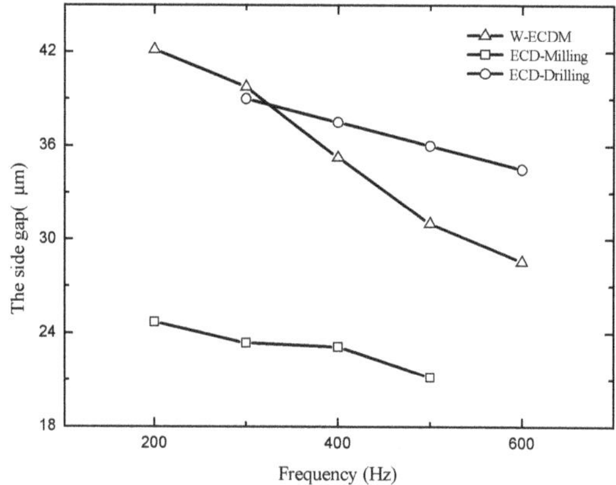

Figure 7. Effect of frequency on side gap.

3.5. *Effect of Feed Rate on Side Gap*

Numerous experiments were conducted to research the effect of feed rate on the side gap. Different feed rates had different influences on the side gap, as displayed in Figure 8. Better machining location with s higher feed rate could be obtained with a lower side gap. The shorter discharge time with the higher feed rate in unit machining distance along the direction of feed, led to less material being removed. Therefore, the side gap was lower in the electrochemical discharge drilling, electrochemical discharge milling, and wire ECDM. However, the optimal feed rate was not higher. The rotary helical electrode collided with the workpiece when the feed rate rose to critical values.

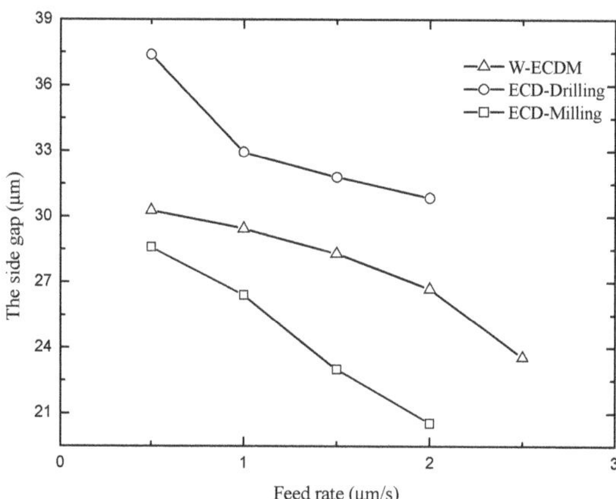

Figure 8. Effect of feed rate on side gap.

4. Experimental Results

According to the above experiments and analysis, the effects of the parameters, including voltage, frequency, duty factor and feed rate, on the side gap were worked out in ECDM with rotary helical electrodes. The parameters, after optimization, were selected based on many experiments exploring fabricated micro holes, micro grooves, micro channels and complicated three-dimensional features with lower side gaps. There were some micro structures displayed.

4.1. Electrochemical Discharge Drilling of Array Micro Holes

According to the above discussion about the effect of the parameters on the side gap, the smaller side gaps needed a low voltage, low duty factor, high frequency, and high feed rate. However, considering material removal rate and machining stability, the experiments were carried out to select a set of optimized parameters for electrochemical discharge drilling. The optimized parameters were: pulse voltage—37 V, frequency—3000 Hz, duty factor—70%, feed rate—1 μm/s, spindle speed—3000 rpm, and electrolyte—3 mol/L KOH. High quality array micro holes were successfully fabricated with a lower diameter, as shown in Figure 9. Thickness of the glass was 300 μm. A minimum side gap of 27.2 μm could be obtained with electrochemical discharge drilling.

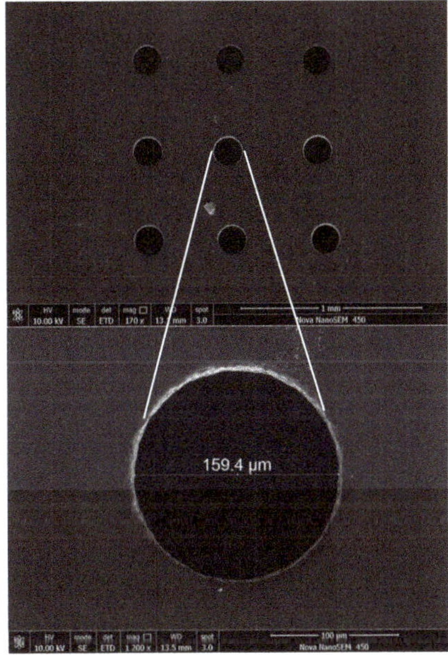

Figure 9. Array micro holes and partial magnification

4.2. Electrochemical Discharge Milling of Micro Structures

Electrochemical discharge milling was capable of fabricating micro grooves, micro channels and micro three-dimensional structures. Some complex micro structures could be machined with a lower side gap by a set of optimized parameters. The optimized parameters were: pulse voltage—34 V, frequency—500 Hz, duty factor—50%, feed rate—2 μm/s, spindle speed—3000 rpm, and electrolyte—3 mol/L KOH. As shown in Figure 10, the micro groove array was milled on the glass. The mean width was 129.4 μm, the length was 750 μm and depth was about 130 μm. The smallest side gap reached 11.5 μm.

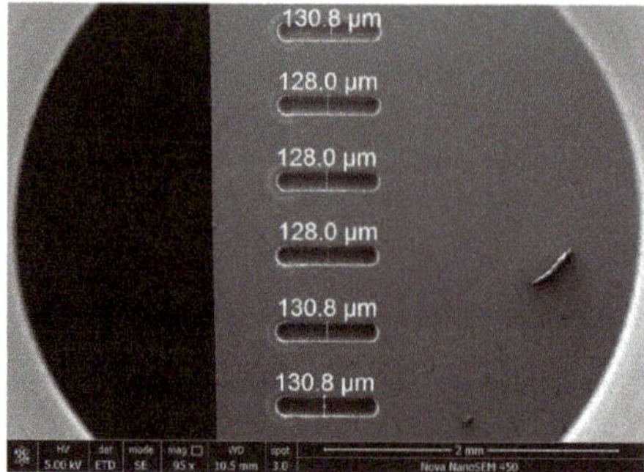

Figure 10. Micro groove array on glass.

The micro channel machined on the glass by electrochemical discharge milling is displayed in Figure 11. The groove width was about 135.9 μm and the depth was about 150 μm. The abbreviation of the university name milled on the glass is shown in Figure 12. The three-dimensional step structure with vertical sidewalls and high shape accuracy is shown Figure 13. The three-dimensional convex structure of micro electrochemical discharge milling is shown in Figure 14, which is made of two layers of convex structures. The width of the upper convex plate was about 75 μm, the length was 260 μm and the height was about 70 μm.

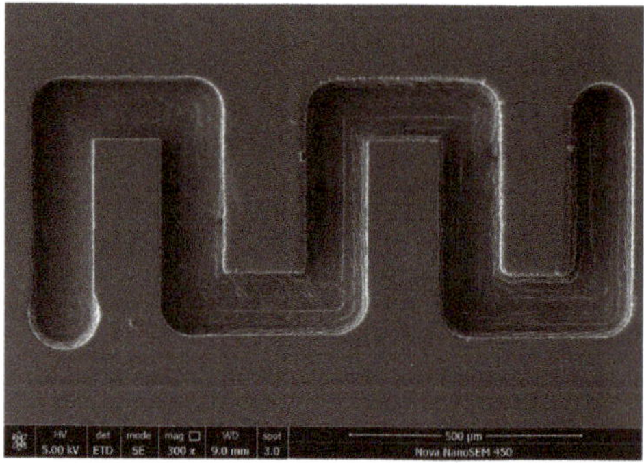

Figure 11. Complex micro channel on glass.

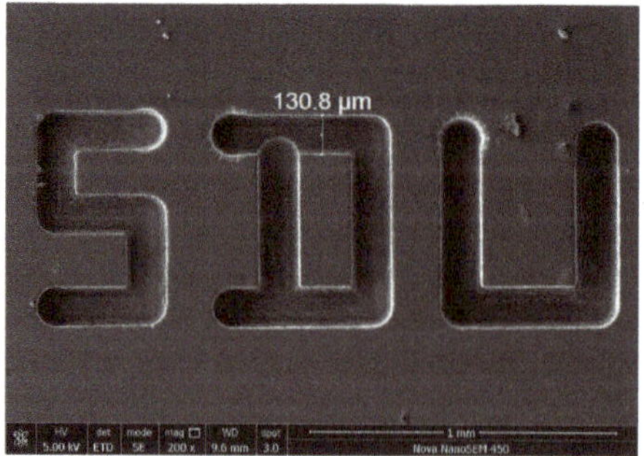

Figure 12. The abbreviation of the university name on glass.

Figure 13. Three-dimensional step structure.

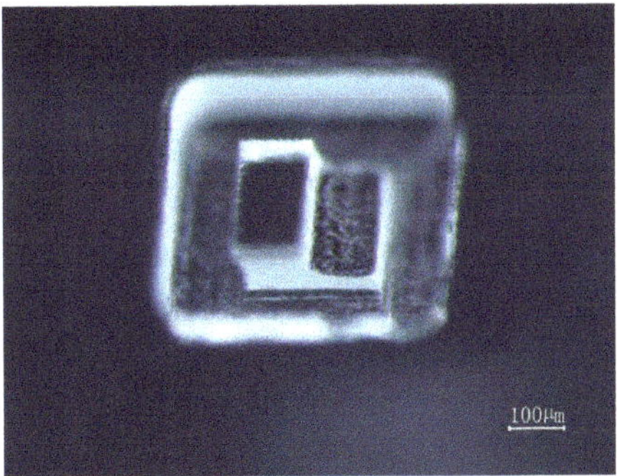

Figure 14. Three-dimensional convex structure.

4.3. Wire Electrochemical Discharge of Micro Structures

Wire electrochemical discharge with rotary helical electrodes fabricated high aspect ratio structures. According to the above discussion and experiments, a set of optimized parameters was selected for machining the micro structures. Long narrow slits were fabricated on 300 μm thick glass, as shown in Figure 15. The smallest side gap reached 14.95 μm. The optimized parameters were: pulse voltage—34 V, frequency—600 Hz, duty factor—50%, and feed rate—1 μm/s. The closed micro structures were machined as displayed in Figure 16. To improve the refreshment of the electrolyte in the closed micro structures, larger processing parameters were used (40 V, 500 Hz, 50%, 1 μm/s).

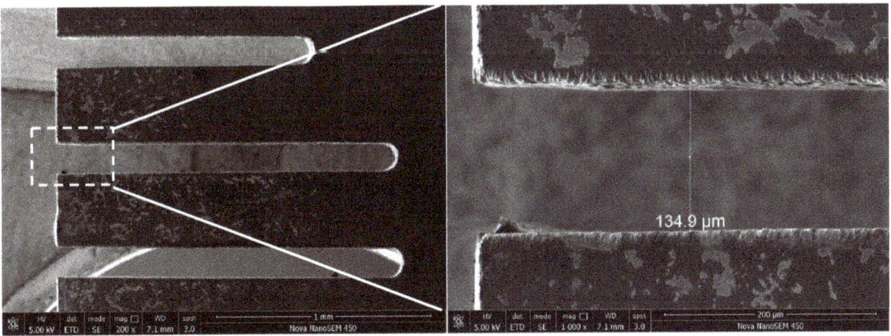

Figure 15. Long narrow slits on 300 μm thick glass workpiece.

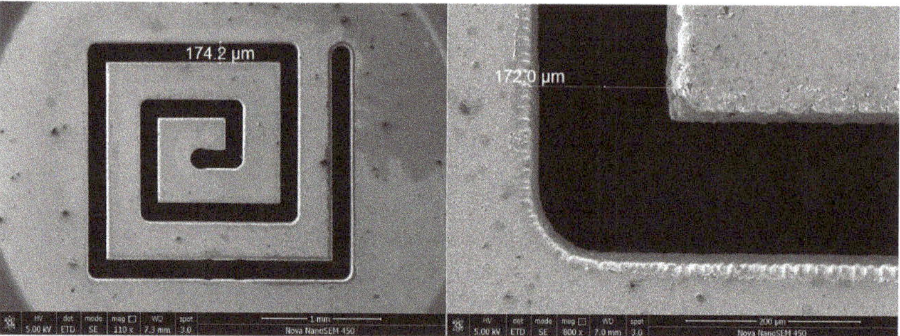

Figure 16. Closed micro structure on 300 μm thick glass.

The high aspect ratio structure was manufactured on 1060 μm thick glass with wire ECDM, using a rotary helical electrode. The slit width was about 175.4 μm and the side gap was about 35.2 μm, as shown in Figure 17 (40 V, 300 Hz, 60%, 1 μm/s). In addition, the micro cantilever beam was successfully fabricated on a 35 μm thick glass workpiece, as shown in Figure 18. The length of the micro cantilever beam was 1500 μm and the aspect ratio reached 42:1.

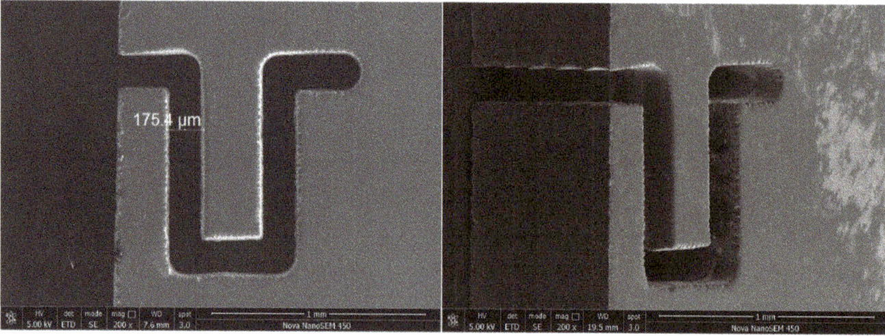

Figure 17. High aspect ratio structure on 1060 μm thick glass.

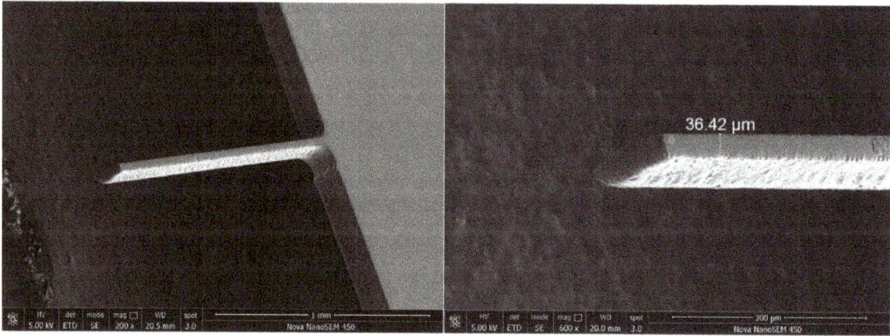

Figure 18. Micro cantilever beam on 35 μm thick glass workpiece.

5. Conclusions

This research employed ECDM with a rotary helical electrode to fabricate ultra-clear glass. Using a rotary helical tool in electrochemical discharge drilling, electrochemical discharge milling, and wire ECDM, the effects of pulse voltage, frequency, duty factor, and feed rate on the side gap were investigated. The conclusions can be summarized as follows:

(1) The mathematical model for the ECDM process was established to guide the machining of microstructures on ultra-clear glass.
(2) The side gap increased with the increase in voltage and duty factor and was reduced with a higher frequency and feed rate in a certain range.
(3) By employing optimized parameters in ECDM, micro holes, micro channels, micro slits and complicated three-dimensional features with ten several-micron side gaps were successfully fabricated on ultra-clear glass.

Author Contributions: Conceptualization, Y.L. and C.Z.; investigation, S.L. and Z.W.; methodology, C.G.; project administration, Y.L.; resources, Z.W.; writing—original draft, S.L.; and writing—review and editing, Y.L. and C.G.

Acknowledgments: The authors acknowledge financial support from the Shandong Provincial Natural Science Foundation (Nos. ZR2018MEE018, ZR2017BEE012), the China Postdoctoral Science Foundation (No. 2018M630772), the Young Scholars Program of Shandong University, Weihai (2015WHWLJH03), and the Shenzhen Science and Technology Project (JCYJ20170818103826176).

Conflicts of Interest: The authors declare no conflicts of interest.

References

1. Allesu, K.; Ghosh, A.; Wuju, M.K. A preliminary qualitative approach of a proposed mechanism of material removal in electrical machining of glass. *Eur. J. Mech. Eng.* **1991**, *36*, 201–207.
2. Zheng, Z.P.; Cheng, W.H.; Huang, F.Y.; Yan, B. 3D microstructuring of Pyrex glass using the electrochemical discharge machining process. *J. Micromech. Microeng.* **2007**, *17*, 960–966. [CrossRef]
3. Díaz-Tena, E.; Gallastegui, E.; Hipperdinger, M.; Donati, E.R.; Ramírez, M. New advances in copper biomachining by iron-oxidizing bacteria. *Corros. Sci.* **2016**, *112*, 385–392. [CrossRef]
4. Paredes-Sanchez, J.P.; Lopez-Ochoa, L.M.; Lopez-Gonzalez, L.M.; Las-Heras-Casas, J.; Xiberta-Bernat, J. Evolution and perspectives of the bioenergy applications in Spain. *J. Clean. Prod.* **2019**, *213*, 553–568. [CrossRef]
5. Sanchez, J.A.; Plaza, S.; De Lacalle, L.N.; Lamikiz, A. Computer simulation of wire-EDM taper-cutting. *Int. J. Comput. Integr. Manuf.* **2006**, *19*, 727–735. [CrossRef]
6. Nasim, S.; Mohammad, R.R.; Mansour, H. Experimental investigation of surfactant-mixed electrolyte into electro chemical discharge machining (ECDM) process. *J. Mater. Process. Technol.* **2017**, *250*, 190–202.
7. Zhang, Z.Y.; Huang, L.; Jiang, Y.J.; Liu, G.; Nie, X.; Lu, H.Q.; Zhuang, H.W. A study to explore the properties of electrochemical discharge effect based on pulse power supply. *Int. J. Adv. Manuf. Technol.* **2016**, *85*, 2107–2114. [CrossRef]
8. Sathisha, N.; Somashekhar, S.H.; Shivakumar, J. Prediction of material removal Rrate using regression analysis and artificial neural network of ECDM process. *Int. J. Recent Adv. Mech. Eng.* **2014**, *3*, 69–81.
9. Jawalkar, C.S.; Sharma, A.P.; Kumar, P. Micromachining with ECDM: Research Potentials and Experimental Investigations. *Channels* **2012**, *6*, 340–345.
10. Cao, X.D.; Kim, B.H.; Chu, C.N. Hybrid Micromachining of Glass Using ECDM and Micro Grinding. *Int. J. Precis. Eng. Manuf.* **2016**, *11*, 5–10. [CrossRef]
11. Elhami, S.; Razfar, M.R. Effect of ultrasonic vibration on the single discharge of electrochemical discharge machining. *Mater. Manuf. Process.* **2017**, *33*, 444–451. [CrossRef]
12. Elhami, S.; Razfar, M.R. Analytical and experimental study on the integration of ultrasonicallyvibrated tool into the micro electro-chemical discharge drilling. *Precis. Eng.* **2017**, *47*, 424–433. [CrossRef]
13. Singh, T.; Dvivedi, A. Developments in electrochemical discharge machining: A review on electrochemical discharge machining, process variants and their hybrid methods. *Int. J. Mach. Tools Manuf.* **2016**, *105*. [CrossRef]
14. Elhami, S.; Razfar, M.R. Study of the current signal and material removal during ultrasonic-assisted electro chemical discharge machining. *Int. J. Adv. Manuf. Technol.* **2017**, *92*, 1591–1599. [CrossRef]
15. Han, M.S.; Min, B.K.; Lee, S.J. Modeling gas film formation in electrochemical discharge machining processes using a side-insulated electrode. *J. Micromech. Microeng.* **2008**, *18*, 19–26. [CrossRef]
16. Furutani, K.; Maeda, H. Machining a glass rod with a lathe-type electro-chemical discharge machine. *J. Micromech. Microeng.* **2008**, *18*, 6–13. [CrossRef]

17. Kun, L.W.; Hsin, M.L.; Kuan, H.C. Application of Electrochemical Discharge Machining to Micro-Machining of Quartz. *Adv. Mater. Res.* **2004**, *939*, 161–168.
18. Tsuchiya, H.; Inoue, T.; Miyazaiki, M. Wire electro-chemical discharge machining of glasses and ceramics. *Bull. Jpn. Soc. Precis. Eng.* **1985**, *19*, 73–74.
19. Jain, V.K.; Rao, P.S.; Choudhary, S.K.; Rajurkar, K.P. Experimental Investigations into Traveling Wire Electrochemical Spark Machining (TW-ECSM) of Composites. *J. Eng. Ind.* **1991**, *113*, 75–84. [CrossRef]
20. Panda, M.C.; Yadava, V. Finite element prediction of material removal rate due to traveling wire electrochemical spark machining. *Int. J. Adv. Manuf. Technol.* **2009**, *45*, 506–520. [CrossRef]
21. Kuo, K.Y.; Wu, K.L.; Yang, C.K.; Yan, B.H. Wire electrochemical discharge machining (WECDM) of quartz glass with titrated electrolyte flow. *Int. J. Mach. Tools Manuf.* **2013**, *72*, 50–57. [CrossRef]
22. Wang, J.; Guo, Y.B.; Fu, C.; Jia, Z.X. Surface integrity of alumina machined by electrochemical discharge assisted diamond wire sawing. *J. Manuf. Process.* **2018**, *31*, 96–102. [CrossRef]
23. Rattan, N.; Mulik, R.S. Improvement in material removal rate (MRR) using magnetic field in TW-ECSM process. *Mater. Manuf. Process.* **2017**, *32*, 101–107. [CrossRef]
24. Huang, S.F.; Liu, Y.; Li, J.; Hu, H.X.; Sun, L.Y. Electrochemical Discharge Machining Micro-Hole in Stainless Steel with Tool Electrode High-Speed Rotating. *Mater. Manuf. Process.* **2014**, *29*, 634–637. [CrossRef]
25. Han, M.S.; Min, B.K.; Lee, S.J. Geometric improvement of electrochemical discharge micro-drilling using an ultrasonic-vibrated electrolyte. *J. Micromech. Microeng.* **2009**, *19*, 65004. [CrossRef]
26. Fang, X.L.; Zhang, P.F.; Zeng, Y.B.; Qu, N.S.; Zhu, D. Enhancement of performance of wire electrochemical micromachining using a rotary helical electrode. *J. Mater. Process. Technol.* **2016**, *227*, 129–137.
27. Wang, M.Y.; Zhang, J.H.; Liu, Y.; Li, M.H. Investigation of Micro Electrochemical Discharge Machining Tool with High Efficiency. *Recent Pat. Eng.* **2016**, *10*, 146–153. [CrossRef]
28. Jain, V.K.; Dixit, P.M.; Pandey, P.M. On the analysis of the electrochemical spark machining process. *Int. J. Mach. Tools Manuf.* **1999**, *39*, 165–186. [CrossRef]

MDPI
St. Alban-Anlage 66
4052 Basel
Switzerland
Tel. +41 61 683 77 34
Fax +41 61 302 89 18
www.mdpi.com

Processes Editorial Office
E-mail: processes@mdpi.com
www.mdpi.com/journal/processes

www.ingramcontent.com/pod-product-compliance
Lightning Source LLC
LaVergne TN
LVHW070906170726
843515LV00005B/719

* 9 7 8 3 0 3 9 3 6 3 8 6 5 *